THE AMAZON

WHAT EVERYONE NEEDS TO KNOW®

MARK J. PLOTKIN

OXFORD
UNIVERSITY PRESS

OXFORD
UNIVERSITY PRESS

Oxford University Press is a department of the University of Oxford. It furthers the University's objective of excellence in research, scholarship, and education by publishing worldwide. Oxford is a registered trade mark of Oxford University Press in the UK and certain other countries.

"What Everyone Needs to Know" is a registered trademark of Oxford University Press.

Published in the United States of America by Oxford University Press
198 Madison Avenue, New York, NY 10016, United States of America.

Library of Congress Cataloging-in-Publication Data
Names: Plotkin, Mark J., author.
Title: The Amazon : what everyone needs to know / Mark J. Plotkin.
Description: New York : Oxford University Press, [2020] |
Series: What everyone needs to know |
Includes bibliographical references and index.
Identifiers: LCCN 2019052408 (print) | LCCN 2019052409 (ebook) |
ISBN 9780190668297 (hardback) | ISBN 9780190668280 (paperback) |
ISBN 9780190668310 (epub)
Subjects: LCSH: Rain forest ecology—Amazon River Region. |
Biodiversity—Amazon River Region. | Deforestation—Amazon River Region. | Human ecology--Amazon River Region.
Classification: LCC QH112 .P56 2020 (print) | LCC QH112 (ebook) |
DDC 577.340981/1--dc23
LC record available at https://lccn.loc.gov/2019052408
LC ebook record available at https://lccn.loc.gov/2019052409

1 3 5 7 9 8 6 4 2

Paperback printed by LSC Communications, United States of America
Hardback printed by Bridgeport National Bindery, Inc., United States of America

I dedicate this book to my mother, Helene T. Plotkin,
who taught me to love and respect Mother Nature.

We rush around like the demented: in the first three days, we were unable to classify anything; we pick up one object to throw it away for the next. . . .

[My colleague] Bonpland keeps telling me he will go mad if the wonders do not cease. . . .

—Alexander von Humboldt, describing his first forays into the Amazon rainforest in a letter home (1799)

CONTENTS

ACKNOWLEDGMENTS

The author wishes to thank Alexandra Aikhenvald, Amasina Alalapadu, Marie Arana, Daniel Aristizabal, Bruce Babbitt, Henrik Balslev, Arianna Bastos, Adam Bauer-Goulden, Tim Bent, Rhett Butler, Janell Cannon, Jim Castner, Ken Catania, Sandra Charity, Charles Clement, Jack Cover, Tom Dillehay, Philip Fearnside, Adrian Forsyth, Carolina Gil, Jeffrey Gorham, Peter Gorman, Michael Goulding, Greg Grandin, John Hemming, Andrew Henderson, Brian Hettler, Bruce Hoffman, Carina Hoorn, Rudo Kemper, Zack Lemann, Tom Lovejoy, John Lundberg, Liliana Madrigal, Dennis McKenna, German Mejia, Santiago Mora, Mary O'Grady, Enrique Ortiz, Kamainja Panashekung, Minu Parahoe, Francelys Peche, Antonio Peluso, Dominiek Plouvier, Petru Popescu, Anthony Rylands, Rafe Sagalyn, Neil Schultes, Mike Shanahan, Glenn Shepard, Deborah Snyder, David Stone, John Terborgh, Michael Totten, Daniel Tredigdo, Kirk Winemiller, and Abigail Wright.

Map 1 Map of northern South America with the shaded area showing the distribution of the Amazon rainforest (Map by Brian Hettler and Rudo Kemper of the Amazon Conservation Team).

Map 2 Map of northern South America featuring locales mentioned in the text (Map by Brian Hettler and Rudo Kemper of the Amazon Conservation Team).

1

INTRODUCTION

What is a tropical rainforest?

Rainforests occupy a special place in the imagination. Literary, historical, and cinematic depictions range from a ghastly green hell to an idyllic Garden of Eden. In terms of fiction, they fired the already fervent imaginations of storytellers as diverse as Sir Arthur Conan Doyle, Edgar Rice Burroughs, Rudyard Kipling, and even George Lucas and Steven Spielberg in whose books and films they are inhabited by dinosaurs, trod by Indiana Jones, prowled by Mowgli the Jungle Boy, and swung through by Tarzan of the Apes.

But rainforest fact is no less fascinating than rainforest fiction. Brimming with mystery and intrigue, these forests still harbor lost cities, uncontacted tribes, ancient shamans, and powerful plants than can kill and cure.

The rainforest bestiary extends far beyond the requisite lions, tigers, and bears. Flying foxes and winged lizards, arboreal anteaters, rainforest giraffes, cross-dressing spiders that disguise themselves as ants, and bats the size of a bumblebees all flourish in these most fabulous of forests, along with other zoological denizens that are equally bizarre and spectacular. And no scientist immersed in these ecosystems believes that all the wonders have been found or revealed.

Tropical rainforests merit their moniker. They flourish in the tropics—the more than 3,000-mile-wide (4,830 km) equatorial band between the Tropic of Cancer and the Tropic of Capricorn. And these forests are hot, humid, and wet, receiving, in the Amazon, on average from 60 to 120 inches (152–300 cm) of rain per year, as compared to a mere 25 inches (63 cm) in London or 45 inches (114 cm) in Manhattan. However, several sites in the rainforests of northeastern India, western Africa, and western Colombia are drenched by more than 400 inches (10 m) of precipitation per annum.

To a large degree, rainfall in the tropics is determined by the so-called *Intertropical Convergence Zone* (ICZ), a band of clouds around the equator created by the meeting of the northeast and southeast trade winds. This band is also referred to as the "Monsoon Trough" and known to—and dreaded by—sailors over the centuries as the "Doldrums," since the extended periods of calm that sometimes manifest there could strand a sailing vessel for weeks.

The constant cloud cover due to the ICZ, the ferocious heat, and the abundant rainfall combine to produce high humidity, sometimes close to 95% in the Amazon, a challenge for visitors unused to such torpor. According to Rhett Butler of Mongabay: "Each canopy tree transpires 200 gallons (760 L) of water annually, translating roughly into 20,000 gallons (76,000 L) transpired into the atmosphere for every acre of canopy trees. Large rainforests (and their humidity) contribute to the formation of rain clouds, and generate as much as 75% of their own rain and are therefore responsible for creating as much as 50% of their own precipitation."

Temperate rainforests, by comparison, typically occur in coastal strips in temperate regions, like the coniferous forests found from Oregon to Alaska. Other temperate rainforests flourish in Australia, Japan, Russia, and New Zealand. Whereas tropical rainforest trees are often bedecked with vines and orchids, these temperate forests are typically festooned with ferns and mosses.

Conservationist Thomas Lovejoy calls tropical rainforests "the greatest expression of life on earth." Unlike the gymnosperms (conifers) that dominate the temperate rainforests, the tropical rainforests consist mostly of evergreen woody angiosperms (flowering plants). A peculiar and unforgettable aspect of these equatorial rainforests is the ubiquity of *lianas*, woody vines that duck and dodge and climb and curl throughout the ecosystem. Palm trees are also emblematic denizens of these same forests; more than three-quarters of the world's 200 palm genera originate in the tropical belt. Global commodities produced by tropical palms include acaí fruit, carnauba car wax, coconuts, palm hearts, palm oil, and rattan furniture.

Tropical rainforest trees often feature *buttress roots*, large above-ground structures that help support the tall but often shallow-rooted trees. These roots flare out from the trunks, not unlike the flying buttresses of medieval cathedrals, and serve a similar function in terms of providing lateral support to a heavy central structure. Buttresses are believed to reduce soil erosion and to trap leaf litter, thereby providing the tree additional nutrients. They increase water uptake and storage and increase the efficiency of nutrient absorption in an ecosystem where most of the nutrients occur on or near the surface. Buttress shape and height—some grow as high as 30 feet (10 m)—are very distinctive and therefore useful in helping botanists identify to which species the tree belongs.

Many leaves of flowering plants—particularly in the forest canopy—are relatively thin and feature "drip tips" at the end. They accelerate water runoff and the drying of the leaf surface, adaptations that inhibit fungi and bacteria from growing on and attacking the leaf. Thicker leaves are more common deeper inside the forest, where less light is available for photosynthesis.

Another common but peculiar phenomenon that typifies rainforest ecosystems is *cauliflory*, in which fruits and flowers sprout directly from the tree trunks or branches rather than at the tips of branches. Such a process may facilitate pollination and/or seed dispersal by large animals or even by smaller creatures that can neither climb nor fly. Obvious and well-known examples in Amazonia include cacao/chocolate (*Theobroma cacao*), the calabash tree (*Crescentia cujete*), and papaya (*Carica papaya*).

Unlike the relatively depauperate temperate ecosystems, the tropical rainforests often feature almost unimaginably high levels of biodiversity; estimates are that they harbor as much as 50% of the world's terrestrial biodiversity on a mere 16% of the earth's surface. The great Harvard entomologist E. O. Wilson found that a single bush in Bolivia may be home to more species of ants than the entire British Isles. A single tropical river may harbor more fish species than all the rivers of Europe combined. The little country of Belize in Central America is smaller than the state of New Hampshire yet has more than twice as many species of bats as found in the entire United States.

And the intimate and interlocking relationships between various species in the tropics are equally stunning. In the temperate zone, for example, most bats prey on insects. In the Amazon, there are bats that feed on insects, spiders, fruit, nectar, seeds, pollen, fish, birds, frogs, lizards, snakes, small rodents, and even other bats. And there are bats that drink blood: bird blood, cow blood, horse blood—and even human blood. In the South American rainforest, not only are there bats that eat frogs but there are at least three species of frogs that devour bats: the smoky jungle frog (*Leptodactylus pentadactylus*), the colorful and bizarre horned frog (*Ceratophrys* spp.), and the giant cane toad (*Rhinella marina*).

In the temperate zone, most pollination is carried out by wind and insects. Inside the rainforest, however, there exists little wind so much of the pollination (and seed dispersal) is carried out by animals, including bats, birds, lizards, monkeys, and rodents. Some Amazonian trees reproduce by dropping their fruits into rivers where they are eaten by fish, which then disperse the seeds far from parent plant, an unusual codependency between aquatic and terrestrial species.

Ecologists recognize several "layers" of the rainforest, each of which has its own characteristics. The "roof" layer of the rainforest is the canopy, formed by the uppermost portions of most of the tallest trees in the rainforest. Because of its height—the canopy can be more than 100 feet (30 m) above the ground and more than 30 feet (10 m) thick—it remains the least studied and most poorly known part of the rainforest. (In recognition of the wonders still to be discovered, one rainforest entomologist hails the canopy as "the high frontier.") The canopy is exceptionally rich in epiphytes—air plants—that comprise "mini-ecosystems" with their own flora and fauna. Overall, the rainforest canopy harbors such unlikely creatures as flying snakes, poison dart frogs, crabs, arboreal anteaters, and even tree kangaroos!

As in the ocean, life in the rainforest is concentrated in the upper layers, where most of the sunlight strikes and is absorbed into the ecosystem. The canopy is the primary site for photosynthesis, pollination, seed dispersal, herbivory, and even biodiversity. As such, canopies have been called "green oceans." The late ecologist William Beebe famously wrote that "another continent of life remains to be discovered, not upon the Earth, but one to two hundred feet above it, extending over thousands of square miles."

Initial attempts to study the canopy from a scientific perspective were inefficient and ineffective: they ranged from felling the individual specimens (no easy task when each tree may be more than 150 feet (45 m) tall and literally tied to its neighbors by numerous lianas) to actually climbing the trees (ditto). Further complicating the research was the prevalence of biting and stinging insects and arachnids, not to mention poisonous snakes. A pioneering approach to canopy research was developed in the 1980s, in Guyana, by the swashbuckling filmmaker Neil Rettig, who employed a crossbow to fire ropes over the tallest trees, then used the ropes to haul wood (and himself) up to the canopy to build tree houses from which he would film and study harpy eagles in situ for months at a time. Scientists now visit and even live in the upper echelons of the world's rainforests employing tools that range from hot air balloons to cranes to canopy walkways constructed through the treetops.

Tropical rainforests canopies are overtopped by so-called *emergents*, individual and isolated giant trees that tower over the rest of the forest. Trees like angelim vermelho (*Dinizia excelsa*) and kapok (*Ceiba pentandra*) in the Amazon and yellow meranti (*Shorea faguetiana*) and tualang (*Koompasia excelsa*) in tropical Asia can exceed 200 feet (60 m) in height and feature crowns up to 70 feet (21 meters) in length. Because they are so exposed, these trees must be able to withstand strong winds, high heat, low humidity, and torrential rain. Due to their isolation in terms of emerging above and beyond the rest of the forest, these trees appear to stand as lonely sentinels watching over the emerald ecosystem below.

Almost every visitor to tropical rainforests notes that—due to the presence of the canopy overhead—little direct sunlight penetrates into the forest. Only about 5% (or less) of the sunlight striking the emergent and canopy layers penetrates deep into the forest. Here, inside the forest, is the so-called *understory* where the lack of light makes the surroundings gloomy but certainly not impenetrable since plant life is thin. One colleague has wryly noted that here a flashlight is perhaps more useful than a machete. In fact, the only really impenetrable part of a rainforest lies along rivers or roads, or when a large tree has fallen and the vegetation that has grown up beside it consists of dense and seemingly impassable tangles of sun-seeking spiny vines and lianas.

Given the canopy overhead, there is little in the way of wind or air circulation in the understory, so humidity is high and temperature remains relatively steady. The vegetation is sparse, often dominated by young individuals of canopy species—"suppressed juveniles"—waiting for a gap to form in the canopy to provide the light that they require to grow to maturity. Scattered shrubs and small palms are not uncommon. Because herbaceous plants of the understory are adapted to conditions of low light and abundant water, some, like *Calathea* and *Maranta*, have become popular houseplants.

With most of what light penetrates through the canopy being absorbed by the understory, even less of it makes it to the forest floor layer, resulting in the presence of very few flowering plants. As a result, many of the small number of herbaceous plants found in the lower reaches of the rainforest—like aroids, ferns, and orchids—survive as epiphytes growing on tree stems or branches where the limited light is more abundant.

The forest floor itself—where the actual decomposition takes place—is dark and humid. Most of the soils are poor, acidic, and heavily eroded due to the constant precipitation. There is also a continuous rain of dead flowers, twigs, and leaves, which are then broken down and absorbed by tiny fungal threads feasting on these end products of photosynthesis. So dominant is the fungal lifestyle in this lightless zone that some flowering plants have evolved into fungal mimics. Rafflesia, for example, is a flowering plant that produces no leaves, stems, or roots and lives by parasitizing vines. Native to the rainforests of Indonesia, the remarkable *Rafflesia arnoldia* produces the world's largest flower: a striking (but odiferous) red and orange beauty, more than 3 feet (1 m) in diameter, and weighing more than 20 pounds (9.5 kilograms). Pollinated by flies, it emits a foul odor said to resemble rotting flesh, earning it the nickname "corpse flower."

An almost equally odd New World counterpart flourishes in the Amazon rainforest. Looking precisely like a red-orange mushroom sprouting from the soil, *Helosis cayennensis* is actually a flowering plant that parasitizes the roots of trees. The indigenous peoples of Suriname, never at a loss for evocative plant names, call this rainforest floor denizen *didribi warung* (the devil's penis).

What do we mean when we say "the Amazon"?

Redolent of mystery, intrigue, danger, and beauty, the Amazon conjures up visions of trackless rainforests, lost tribes, healing plants, furtive headhunters, deadly blowdarts, man-eating fish, lethal serpents, and awe-inspiring waterfalls. It is the embodiment of the beautiful and dangerous. Some films picture it as a green hell while others see it as an emerald Eden. Even among specialists, there exists broad disagreement as to what the Amazon actually is, particularly in terms of borders and boundaries.

Many use the terms "Amazon rainforest" and "Amazon Basin" interchangeably. They are not the same thing. The Amazon River originates in the Andean highlands of southern Peru before passing southeastern Colombia, taking in huge volumes of water from Bolivian and Ecuadorian rivers along its journey and eventually emptying into the Atlantic. The Amazon rainforest, however, is not limited to the Amazon Basin, as it reaches its leafy tendrils into Venezuela (mostly in the Orinoco Basin) as well as into all three Guianas (French Guiana, Guyana, and Suriname),[1] embedded in the northeast shoulder of South America. This entire area has been referred to as "the Amazon biome," "the biogeographic Amazon," "the ecological Amazon," and the more prosaic (and less evocative) "AOG region" (Amazon, Orinoco, and Guianas). For the purposes of this book, I refer to it as "Amazonia." By this more inclusive definition, the Amazon rainforest encompasses a total of 3.2 million square miles (9 million km^2), which is roughly the size of the contiguous 48 US states. Brazil contains most of Amazonia: approximately 60%. Just over 11% of Amazonia is found within the borders of Peru, followed by Colombia (8%), Venezuela (7%), Bolivia (6%), Guyana (3%), Suriname (2%), Ecuador (2%), and French Guiana (1%).

Amazonia consists of more than rainforests because there are extensive aquatic ecosystems (not just rivers, but also lakes, swamps, and seasonally flooded wetlands) as well as savannas, palm forests, and odd scrubby botanical assemblages atop peculiar table mountains. A 19th-century explorer described Amazonia as quite simply the largest river flowing through the greatest rainforest in the world.

Such superlatives are mirrored in the region's biodiversity because Amazonia contains the largest living assemblage of plants and animals on the planet. More than one of every four flowering plant species on earth resides in the Amazon. Manu National Park in southeastern Peru harbors 25% more species of birds than are found in the entire United States and four times more butterflies than inhabit Europe. Amazonia has more types of fish than the Nile, the Congo, and the Mississippi combined. Cantao State Park in southeast Amazonia contains more fish species than all the rivers of Europe combined.

Almost everything about Amazonia seems outsized. It is home to the world's largest anteater, armadillo, eagle, caecilian, freshwater fish, freshwater turtle, snake, and spider. The Amazon also boasts the world's heaviest rodent (which weighs more than 200 lbs /91 kilograms), catfish (250 lbs /114 kilograms), and alligator (more than half a ton /454 kilograms)—and the latter two are known man-eaters.

Amazonia is a biological wonderland, brimming with mind-bendingly bizarre, creepy, weird, and wonderful creatures. Its rainforests and rivers harbor tree-eating catfish, vampire bats and vampire fish, pink dolphins, 4-foot-long (1.2 meter) earthworms, predatory glow worms, monkeys the size of mice, monkey-eating fish, lizards that run on water, sloths that swim, giant phallic legless amphibians, archaeopteryx-like birds, and insects that appear to be healthy leaves, or diseased leaves, or half-eaten leaves, or leaves covered in bird droppings. There are frogs flat as a pancake, frogs that are toxic, and frogs that are hallucinogenic.

And species previously unknown to science are constantly being discovered—one recent estimate calculated that a new species is collected in Amazonia every 3 days. Over the course of the past two decades, scientists have collected more than a thousand new species of plants, more than 400 fish (including two new species of electric eels and two new species of stingrays, as well as other fish species that are transparent), more than 300 amphibians (including new poison dart frogs), more than 65 species of mammals (including seemingly conspicuous creatures like monkeys as well as a new dolphin, peccary, and tapir), and insects and other invertebrates too numerous to count (including a cobalt-blue tarantula). Nor is there any indication that we are anywhere near finding all of the undiscovered species that remain.

For example, estimates by botanist Dr. Hans ter Steege and his colleagues at the Naturalis Biodiversity Centre in Holland estimate that the Amazon is home to 16,000 species of trees. The most common families are Fabaceae (legumes), Rubiaceae (coffee family), Melastomataceae (melastome family), Myrtaceae (myrtle family), Lauraceae (cinnamon family), Annonaceae (soursop family), and the Euphorbiaceae (rubber family). Even though 14,000 tree species have been collected and identified, ter Steege estimates that another 2,000 are yet to be found. Nor is our ignorance limited to the number of species: in 2019, Brazilian scientists found the tallest tree in the Amazon, a specimen of angelim vermelho (*Dinizia excelsa*). At a height of 288 feet (88 m), this forest giant reaches almost 100 feet (30 meters) higher than the previous record holder.

Though more than 10% of Amazonian trees remain unknown to science, the future of these plants seems increasingly imperiled. Thirty years ago, "Save the Rainforest!" was the rallying cry for the global environmental movement. This culminated in the Rio Environmental Conference in 1992, which attracted almost every global head of state. With Brazil serving as conference host, "Save the Amazon!" seemed to be an international goal. The Rio Conference focused on environmental protection and sustainable development. Were it to convene today, the issue of climate change would predominate. The global perception of the Amazon's importance for human well-being has clearly declined.

Many Amazonian shamans—the greatest indigenous naturalists—insist on the interconnectedness of all things. This is demonstrably the case with the Amazon and climate change: the number two cause of carbon being released into the atmosphere is deforestation, much of it in the tropics. Satellites monitoring forest integrity have recorded more than 20,000 fires in a single day within the borders of Brazil. And we know—far better than we did at the time of the Rio Conference—that Amazonia plays a key role in stabilizing both the world's climatological and hydrological cycles while providing food and water security to tens of millions of people in the region.

In the age of accelerating climate change, the fate of Amazonia affects everyone. Home to almost 400 *billion* trees, this rainforest plays a vital role in stabilizing global climate by absorbing titanic amounts of carbon dioxide—or releasing it into the atmosphere if immolated. Living forests therefore mitigate climate change, while deforestation exacerbates the process. Rainforests also exhale enormous amounts of water during photosynthesis. Scientists estimate that most of the moisture in the Amazon remains in this relatively closed cycle as the rain returns the water to the forest and the rivers. Deforestation, however, breaks the cycle—rainfall declines and droughts occur. Severe droughts in both Brazil and Southeast Asia have been linked to this disruption. And the impact of these interruptions of the climatic patterns is not merely local: changing rainfall patterns in the United States, Europe, and China have been directly tied to tropical deforestation. The battle against climate change may not be won in Amazonia, but it can certainly be lost there.

Amazonia currently serves as home to approximately 34 million people. In fact, very few of these millions have much of a say in "development" projects being devised by Chinese loggers, Canadian miners, American agribusiness mavens, and multilateral banks in Washington, London, and even Brasilia.

So that is what we mean we talk about "the Amazon": one of the earth's greatest wonders, knowable and unknowable, enduring and vulnerable, and a place of infinite wonder that is increasingly threatened. We ignore this at our peril.

What makes the Amazon important?

The Amazon River and rainforest play an outsized role in the well-being of the world. Despite covering less than 10% of the planet's surface, this northern half of the South American continent harbors a sizeable fraction of the world's terrestrial and freshwater diversity. As we have seen, this biodiversity is extraordinarily interdependent with reference to food web maintenance, pollination, nutrient cycling, and a host of other ecosystem processes, meaning that the destruction of any one of the millions of species living there could have a disastrous impact on many, many others.

These plant and animal species benefit local, regional, national, and even international populations as well. The most immediate beneficiaries of local biota are local people living in the forest: first and foremost, the Amerindians, some of whom (in the case of uncontacted or isolated tribes) draw 100% of their sustenance from the wild. Close to 400 tribes inhabit the Amazon today, the great majority of which still depend on the forest and the rivers for their livelihood: for food, medicine, shelter, fuel, and many other sundry purposes. And an increasing percentage of Amazon inhabitants are not Amerindians: they are peasants, *mestizos, caboclos*, Afro-Americans, or—frequently—some combination of several of these categories. Of the more than 30 million people inhabiting Amazonia, many depend to some degree on the wilderness around them. Most specialists estimate that people living in or near tropical forests usually derive more than 20% of their household income from nature, largely from timber, non-timber forest products, or fish.

Of course, the value of these plant and animal species does not accrue solely to local forest inhabitants: some yield commercial benefits at the urban, regional, state, national, or even international level. Products made from acaí fruits and Brazil nuts are sold in almost every major American supermarket.

Nor is the current or potential medicinal value of forest denizens relevant only to locals: species like jaborandi and plants yielding curare were once exported to meet global demand, while synthetic angiotensin-converting enzyme (ACE) inhibitors—originally derived from the venom of a Brazilian viper—are worth billions of dollars on an annual basis. And other poisons from Amazonian plants and animals—particularly frogs, snakes, insects, and arachnids—are attracting increased research interest. Meanwhile, the treasure trove of flora and fauna known to tribal shamans has never received the investigative attention that their commercial and medicinal promise merits. Further investigations in this field will undoubtedly both further demonstrate and augment the global value of Amazonian biodiversity to human welfare.

Moreover, the agricultural value of Amazonian plants is often poorly recognized. Avocado (*Persea americana*), chile pepper (*Capsicum* spp.), cacao/chocolate (*Theobroma cacao*), manioc (*Manihot* spp.), papaya (*Carica papaya*), peanuts (*Arachis hypogaea*), pineapple (*Ananas comosus*), and yams (*Dioscorea* and *Ipomoea*) all originate—or have close relatives that originate—in Amazonia. The wild or locally grown varieties have an intrinsic value in their availability to cross with widely cultivated varieties around the world to increase resistance to pests and diseases. The Amazon also serves as home to innumerable wild edible species that could potentially be brought into cultivation to feed the hungry or simply to increase our gustatory spectrum—in fact, geographer Nigel Smith compiled a list of more than 100 tasty species in just one corner of the Peruvian Amazon.

And the Amazon plays a vital role in stabilizing the global climate—in fact, probably no other terrestrial ecosystem exerts more control on the carbon cycle. Recent research has concluded that this great rainforest is composed of approximately 39 billion trees—representing about 16,000 different species—and these trees absorb billions of tons of carbon (in the form of carbon dioxide) on a yearly basis. Scientists at the University of Leeds estimate that the Amazon harbors one-fifth of all the planet's terrestrial carbon. Destruction of these forests releases this carbon into the atmosphere, a major driver of climate change. In short, living rainforests absorb and store carbon while deforestation releases it.

These forests also play key roles in local, regional, and global water cycles. The Amazon rainforest is both a sponge and a spout—the ecosystem is profoundly efficient at soaking up the abundant rainfall and then releasing the moisture back into the atmosphere during transpiration. Scientists in Brazil estimate that as much as 80% of the moisture in Amazonia remains in what is essentially a closed cycle. Destroying the forest not only reduces local rainfall—it can lead directly to droughts. Rainfall originating in Amazonia also waters agriculture outside the immediate region, like in the adjacent *cerrado*, Brazil's primary soy-producing region. Deforestation in Amazonia is believed to have contributed to severe drought in São Paulo state—the economic engine of Brazil—and even affects rainfall patterns as far away as the United States.

One of the rainiest places on earth, the Amazon forests anchor everything in place. Unlike the island of Madagascar, where massive deforestation turns rivers red from erosion, the canopy of the standing forest in South America reduces the force of the tropical downpours while the roots bind the soils. However, the staggering devastation of Hurricane Mitch in Central America augurs a much less rosy future for Amazonia if deforestation continues unabated.

In late 1998, Hurricane Mitch smashed into Central America, leaving an unparalleled path of destruction in its wake. Deforested hillsides dissolved into seemingly unstoppable mudslides whose destructive power was magnified many times over by the lack of tropical trees that once secured the soil in place. More than 10,000 people died, and millions were left homeless. The economic damage was estimated to be more than $5 billion. A Category 5 hurricane would have been destructive in any case, but there can be no doubt that the economic cost, the number of people displaced, and the number killed would have been much lower if the forests had been better managed and protected. And the rapidly accelerating pace of global climate change makes more major storms increasingly likely.

Nor are deforestation catastrophes linked only to storms and floods. In 2005, a longstanding drought in the Brazilian state of Acre led to vast areas of burning forest releasing so much smoke into the air that the state's major hospital was overwhelmed with patients suffering from respiratory ailments and eye problems. The most important airport in the state capital was closed for days due to thick smoke reducing visibility to a fraction of normal, incurring huge economic costs. These conflagrations foreshadowed greater infernos to come: more than 70,000 fires were recorded in Amazonia in 2019.

Yet another value that rainforests possess is as ecotourism venues. Tourism represents one of the world's leading industries and is expected to continue to grow and expand. Costa Rica's leading source of foreign income is ecotourism, and there is no reason that many Amazonian countries cannot equal or surpass the amount of income this tiny Central American country is attracting. In an increasingly urbanized world, the hunger for and the value of authentic wilderness experiences can only increase.

Undoubtedly, the most important reason for protecting Amazonian species is ultimately an ethical and a moral one: we must decide whether we want to live in a world—or have our children and grandchildren grow up in a world—that lacks pink dolphins and blue Morpho butterflies because they were driven to extinction through human shortsightedness, stupidity, and greed.

The final reason for protecting and carefully stewarding the Amazon ecosystem is a utilitarian one: biomimicry, the design and production of materials and structures based on models in nature. Velcro—based on burrs that loosely adhere to natural surfaces—is the best-known example. Though Velcro and the recognition of biomimicry as a valid discipline are both of recent vintage, the realization that Mother Nature was the ultimate design and engineering genius was put forward by an Italian polymath more than 500 years ago: "Though human ingenuity may make various inventions which, by the help of various machines answering the same end, it will never devise any inventions more beautiful, nor more simple, nor more to the purpose than Nature does: because in her inventions, nothing is wanting and nothing is superfluous." So said Leonardo da Vinci.

As home to the greatest diversity of terrestrial species on the planet, the Amazon (sometimes literally) oozes potential. Compounds in vampire bat saliva are being studied as models for new anticoagulants, and peptides in frog slime are being investigated as models for new antibiotics. Dart frogs excrete poisonous compounds, and their method for doing so may teach us how to more efficiently de-ice plane wings in inclement weather. Scorpion and snake poisons are instructing us in how to better attack brain cancers and other tumors, while the diffraction of light from the iridescent blue wings of the Morpho butterfly provides hints for the creation of counterfeit-proof currency. In Suriname, the locals have a saying: "The Amazon holds answers to questions we have yet to ask!"

2

GEOLOGY, SOILS, AND VEGETATION

What is the geological history of Amazonia?

The supercontinent Gondwana began breaking up approximately 180 million years ago. South America and Africa started to drift apart about 40 million years after this as the southern Atlantic Ocean formed between them. A highland arose on the eastern edge of South America, causing water and sediment to flow west. Meanwhile, the South American continental plate moved westward.

The South American continental core consisted of a *craton*, a relatively stable slab of ancient Precambrian crust. Around 150 million years ago, during the Mesozoic era, an intracratonic depression formed, separating the north half of the craton (now known as the Guayana Shield) and the south (now known as the Brazil Shield). The crease between them became the future course of today's Amazon River.

Also during the Miocene, some believe that a small northeast-to-southwest–oriented mountain chain known as the Purus Arch uplifted in the middle of the continent: waters east of the arch flowed into the Atlantic Ocean, and waters to the west flowed toward the Pacific.

As early as 140 million years ago, the South American continental plate began colliding with and grinding against the oceanic crust (the "Nazca Plate") under the Pacific Ocean. As a result, the Nazca Plate was forced down—through a process known as *subduction*—and the South American Plate overtopped it. Thus, the Andes began their rise. These mountains trapped the westward flow of humid air from the Atlantic, increasing both rainfall and soil erosion from the nutrient-rich slopes of the eastern Andes. There was also a north-south–oriented foreland basin that carried the water of a paleo-Amazon-Orinoco north to the Caribbean. By 40 million years ago, the Andes had risen high enough that the river's westward flow was cut off, forming a giant lake and trapping marine species like dolphins and stingrays inside, the descendants of whom evolved to survive in what is today the freshwater environment of the Amazon.

According to some authorities, however, the Caribbean/Atlantic intruded into the western Amazon, forming a brackish or freshwater lake as far south as Bolivia. Once the continued Andean uplift closed off the connection to the sea to the north, rains in western Amazonia created a giant freshwater wetland.

As the South American Plate continued to overtop the Nazca Plate—an ongoing process—the Andes kept rising, forming the longest continental mountain chain in the world and the second tallest, exceeded in height only by the Himalayas. Eventually, the Andes rose high enough to tilt the Amazon basin eastward, and the great river began draining into the Atlantic.

Do large and diverse Amazonian rainforests thrive on poor soils?

Textbooks used to proclaim that the tropics would one day serve as the breadbaskets that would feed a burgeoning global population. The reasoning: soils from which great forests have sprouted must be sufficiently deep, rich, and fecund to support intensive agriculture. Such has seldom proved to be the case in most of tropical South America.

Amazonian soils tend to be ancient, weathered, and poor in minerals like calcium, magnesium, phosphorus, and potassium that are vital for plant life. The Brazilian and Guayana Shields—which comprise much of eastern Amazonia—contain some of the oldest exposed surfaces in South America, with several dating back more than 3.5 billion years. In these environments, the nutrients lost to eons of leaching cannot be easily replaced through the weathering of subsoil, a process characteristic of temperate zones. Because many Amazonian soils feature high concentrations of aluminum and hydrogen that attach to chemicals that would otherwise bond with nutrients, they have a very limited capacity to retain otherwise beneficial nutrients. Furthermore, the Amazon has experienced no geologically recent volcanic activity that would transport additional nutrients into the system. This nutrient-poor nature of the extant soil places limits on plant productivity.

Furthermore, the high rainfall and resultant constant moisture that characterize Amazonia encourage the formation of aluminum oxide and iron oxide, often resulting in soils bearing a characteristic reddish color. Not only are these compounds toxic in large amounts, but they also harbor high clay content, reducing their ability to absorb water and thus making it all the more surprising that great forests emerge from these soils.

However, not all tropical soils are poor: volcanic ones can be exceedingly rich. Harvard biologist Peter Ashton has noted that the island of Java—with its fabulously fertile volcanic soils—is inhabited by more than 140 million people in an area only 30% larger than that of Panama, which hosts only 4 million.

Oxisols (sometimes called *ferralsols*) represent the most common soil type in Amazonia, particularly in the east on both the Brazilian and Guayana Shields. One of the most highly weathered and nutrient-poor soils in South America, oxisols harbor high concentrations of aluminum oxides and iron, which gives them a characteristic red or yellow color. Litter composition rates on top of oxisols tends to be high, and oxisols tend to be most common in regions where there is marked seasonality.

Also abundant in Amazonia are the *ultisols*, which—like the oxisols—are low in nutrients due to millions of years of tropical weathering. They are most often found in areas of low seasonality and high rainfall. In fact, oxisols and ultisols together are believed to make up more than half the soils in Amazonia. And the most intriguing of South American rainforest soils are the famed *terra preta do Indio*—"the black dirt of the Indians"—highly fecund black soils well suited for agriculture that were created by the pre-Columbian inhabitants of Amazonia.

What is terra preta?

The most intriguing of South American rainforest soils are the famed *terra preta do Indio*—"the black dirt of the Indians"—highly fecund and well suited for agriculture, which were created by the pre-Columbian inhabitants of Amazonia. Historically, these rich agricultural lands along the Amazon made possible the enormous villages observed by Orellana and Carvajal in 1541.

Terra preta consists of islands of highly fertile soil surrounded by a sea of depauperate soils (mostly oxisols and ultisols, as noted earlier) that characterize most of Amazonia. The soil's characteristic black color derives from a high proportion of charcoal. It contains copious amounts of pottery shards, carbon and ash from both cooking and agricultural fires, turtle shells, and the bones of both humans and wildlife. Terra preta is richer in calcium, nitrogen, phosphorus, and sulfur than the dirt around it, and it harbors a rich microflora—particularly, mycorrhizal fungi—that encourages plant growth, further increasing fertility.[1]

For cultivators, terra preta offers many advantages over other more depauperate Amazonian soils. Not only is it rich in nutrients, but the soil's charcoal binds and retains minerals so it is not prone to nutrient leaching as might otherwise be the case in regions continuously subjected to tropical downpours.

Today, cash crops like black pepper (*Piper nigrum*), mango (*Mangifera indica*), and papaya (*Carica papaya*) are often cultivated on terra preta seemingly without requiring additional fertilizers. As such, locals seek out terra preta sites throughout Amazonia to plant their crops and sometimes even go so far as to dig it up and sell it as potting soil. Some expatriate American Confederates who fled the United States after the Civil War migrated to the Brazilian Amazon, where they farmed terra preta sites on riverine bluffs near Santarém.

Many terra preta sites are linear in shape, situated on bluffs along river courses. They have been found not only in Brazil along the Amazon, but also along the Negro, Tapajos, Tocantins, and Xingu Rivers; in Venezuela along the Orinoco; in Ecuador along the Napo; in Peru along the Ucayali; in Colombia along the Caqueta and the Vaupes; and in Guyana and Suriname along the Courantyne River, which forms the border between the two countries. Several terra preta expanses are vast: the one that underlies most of the city of Santarém covers almost 2 square miles (5 km^2).

Early European settlers in Amazonia were puzzled by the origin of these soils and hypothesized that they may have been produced by Andean volcanoes. We now know that terra preta appears to have initially been created by tribal cultures around 450 AD. Scholars are divided on the degree of human planning involved: some believe that terra preta arose underneath kitchen middens, while others think that these tribal people intentionally smoldered wood and other organic materials in trenches and then swirled this mix into the soil. Regardless, they proved masterful at employing the soil to enhance food production.

Ultimately, a better understanding of the origin and utility of terra preta is not merely an academic or historical exercise: the principles discovered in the study of this soil are serving as the basis of experiments to develop the means to increase food production around the world. Both terra preta and *biochar* sequester carbon effectively, meaning that a better understanding and more widespread use of this Amazonian creation could also theoretically generate a positive effect on climate change.

How does nutrient cycling help lush rainforests flourish?

Rainforests have proved botanically brilliant at capturing and recycling necessary nutrients. Most of the carbon and nutrients are sequestered in the vegetation itself, both living (such as trees and lianas) and dead (such as fallen trees and leaves). In the temperate zone, nutrients released from decaying vegetation percolate into and are stored in the soil to be accessed by plants by means of their root systems, which can penetrate several feet below the surface. By contrast, the rainforest reabsorbs nutrients so rapidly that nutrient-poor soils are the rule rather than the exception.[2]

Rainforests have evolved into such dazzlingly efficient nutrient sponges and trappers that few of these compounds escape into the soil. Because of this, even the tallest trees usually produce relatively shallow roots. Most Amazonian soils and many tree roots are less than a foot (1/3 of a meter) deep, whereas temperate forest roots can descend more than 5 feet (1.5 m). Each year, as much as 20% of the Amazon rainforest biomass (leaves, twigs, branches, boles, etc.) perishes and falls to the forest floor. There the damp, sweltering conditions that prevail promote rapid decomposition by omnipresent bacteria, beetles, fungi, microorganisms, and termites. At the same time, algae, mosses, and lichens filter nutrients out of rainwater that has fallen onto the forest. When these organisms excrete, or when they die, nutrient uptake by surrounding plants can be almost instantaneous.

Rainforests achieve this efficiency in a variety of ways. A primary method is the conservation of nutrients by individual flora: some tropical trees reabsorb nitrogen, phosphorus, and potassium back into the twig before the leaf falls off; others have thick, leathery leaves to deter predation by herbivores; yet others produce compounds that can dissuade would-be predators, for example by gumming up their mouth parts (*Hevea* rubber latex), causing hallucinations (ayahuasca), or through lethal poisoning (curare).

Some rainforest trees do not wait for nutrients to be released into the soil: they produce above-ground roots that interweave into mats that absorb nutrients—literally before they hit the ground. Aerial roots gain further advantage by forming mutualistic partnerships with forest fungi. By colonizing these roots, the fungi acquire a place to live as well as carbohydrates produced by the host tree during photosynthesis, while the tree reaps increased water and nutrient absorption capabilities from the fungi. These mutually beneficial plant–fungi relationships are known as *mycorrhizae*.[3]

Other trees take the above-ground root approach still further: so-called *apogeotropic roots* grow up the stems of neighboring trees. It is believed that these roots absorb nutrients leached out of the canopy during tropical tempests before those nutrients can reach even the above-ground root mats.

Still another rainforest adaptation evolved (at least, in part) to facilitate nutrient uptake are *tree buttresses*. Known also as *buttress roots*, these flat, often plank-like growths on the sides of trees can reach more than 100 feet (30 m) in height and help the tree spread its roots and collect nutrients over a much wider area. As the name implies, they provide additional stability and support—metaphorically, like flying buttresses in ancient cathedrals—by distributing the tree's collective weight more widely, a valuable trait in a tall tree with shallow roots residing in a forest subjected to frequent storms and tropical downpours.

Other means by which the Amazon rainforest nourishes itself are still being studied and documented. Of interest are so-called *wastebasket trees* like the palms of the genus *Astrocaryum* and *Bactris* whose spines help capture dead leaves falling from the canopy, thereby sequestering nutrients that wash down to the trees' own roots as these leaves decompose above the forest floor. Also fascinating are *tank bromeliads*, pineapple relatives residing on the branches of large trees whose leaves capture water and create aquatic ecosystems 100 feet (30 m) above the forest floor. However, the self-contained aspect of Amazonian rainforest is also a liability in today's world: when human intervention aggressively interrupts this finely tuned ecological process, a wasteland can result.

What are the major forest and vegetation types in Amazonia?

Rainforest dominates most of tropical South America, but other vegetation types also flourish there. Factors like climate, geology, river type, soils, topography, and even pre-Columbian human presence play a determinative role in what type of forest occurs in a particular region.

Lowland rainforest—that growing on terra firme, which is (relatively) high, (relatively) dry, non-flooded soil—represents the most extensive forest type, covering approximately 80% of Amazonia. It is taller (with a canopy up to 115 feet [35 m] high, with emergents growing much higher) and more species-diverse (more than 300 tree species per hectare) than all other forest types found in Amazonia. In addition to tall trees, a closed canopy, a sparse herbaceous layer, and large lianas also characterize these forests.

Extensive stands of low biodiversity forest also exist in Amazonia, demonstrating that this ecosystem is more of a mosaic than a single great stand of rainforest. One example is that of an extensive bamboo forest found straddling the border from Madre de Dios in southeastern Peru to Acre state in southwestern Brazil. A similarly extensive ecosystem, but one of low species diversity, is represented by the liana forest growing between the Xingu and Tocantins Rivers in the southeast Amazon.

Savannas—such as the grasslands of the Brazil–Guyana–Suriname border region and the seasonally inundated llanos of the Colombia–Venezuela border—are another major vegetation type, representing about 4% of Amazonia. There are also some stands of dry forest, primarily in the transition area between Amazonia and the so-called *cerrado* savannas to the southeast. These forests feature a moderate amount of biomass and numerous small lianas and lack a closed canopy.

Campinas (also known Amazon *caatingas* in Brazil, *wallabas* in Guyana, and *banas* in Venezuela) flourish on well-drained, nutrient-poor white sands. These ecosystems harbor a forest that is much less tall and less biodiverse than characteristic Amazonian rainforest, typically featuring much-branched trees that are festooned with epiphytes. Campina forests occur throughout Amazonia but are most commonly associated with blackwater river drainages such are those of the Rio Negro of Brazil. A shorthand way of recognizing campinas is to think of them as slightly stunted dense forests growing on white sands.

Floodplain forests—also known as "inundated forests"—play an outsized role in both ecosystem function and local human well-being. In the Amazon Basin, well over 90% of the forest is terra firme, but floodplain forests cover more than 57,000 square miles (148,000 km^2)—an area quite a bit larger than that of England. These forests tend to be smaller in stature (shorter) and support less biodiversity than forests on terra firme.

This brings us to the flooded forests of the Amazon, one of the most unusual and astonishing ecosystems on the planet. These are ecosystems in which rainfall and swollen rivers coalesce to inundate a forest that remains submerged for months at a time. Forest flooding can reach a depth of 50 feet (15 m) near Tefe in the Brazilian Rio Negro, although 40 feet (12 m) is more commonly the deepest point in other seasonally inundated forests. Though most of the floodplain is underwater from 4 to 7 months per annum, Michael Goulding has recorded instances as short as 3 months and as long as 11 months. Among trees of these forests, the javari palm (*Astrocaryum jauari*) holds the record: it can endure having its stem underwater for 340 days per year.

During high water, the flood in the forest provides an unusual vantage point for a visitor. One can canoe through the forest just beneath the canopy, viewing nesting birds, arboreal mammals, birds, reptiles, amphibians, and plants in fruit and flower at eye level as one literally floats through the upper reaches of the Amazon rainforest.

Some plants—particularly palms—have evolved adaptations that permit them to thrive in such waterlogged environments. In addition to the aforementioned javari, other palms such as acaí (*Euterpe oleracea*) and buriti (*Mauritia flexuosa*) have evolved the means to survive in this flooded environment by developed breathing roots known as *pneumatophores*. The much-utilized piassaba fiber palm (*Leopoldinia piassaba*) features special intercellular spaces in its fruit, which presumably facilitates their floating and dispersal. Many trees of flooded forests feature buttress roots, which presumably anchor them with added stability. Nonetheless—given the waterlogged conditions that prevail much of the year—scientists do not yet understand how these trees survive without being able to absorb oxygen through their roots.

Ecologists divided the flooded forest into two basic types: *igapó* and *várzea*. Of the two, igapó is the less common and is caused by flooding from blackwater rivers and swamps. Because there are few nutrients in these ecosystems, igapó forests tend to have lower biodiversity than is found in terra firme and várzea forests.

Várzea forests, on the other hand, abound in rich nutrients because they are bathed in the nutrients that whitewater rivers carry in and deposit from the rich Andean uplands. Much as the Nile did in ancient Egypt, the Andean waters overtop the riverbanks and flood the várzea with nutrient-rich, high-sediment silt, making this the most fertile land in Amazonia. The productivity of the land and adjacent waters ensures that they serve as important breeding grounds for birds, fish, mammals, and reptiles, including many important commercial species. Daly and Mitchell (2000) summed it up memorably: "The floodplains have ecological importance way out of proportion to their area because of their roles in capturing and cycling nutrients, harboring (and feeding) incomparably rich fish life and great invertebrate diversity, stabilizing flooded soils and landscapes, and requiring remarkable physiological adaptations."[4]

Acre for acre, várzea represents the richest and most important fishing habitat in Amazonia, serving as home for some of the most sought-after species, such as the pacu (*Myolossa* sp.), the tambaqui (*Colossoma macropomum*), and the giant arapaima (*Arapaima gigas*). Though innumerable migratory food fishes breed outside the várzea, many of these same species feed there. Numerous species are adapted to feed on fruits and seeds that drop from forest trees; others prey on insects and spiders that fall into the water. According to Goulding, várzea waters serve as an important hiding place in which fish take refuge from commercial fishermen.

Unfortunately, much of the várzea occurs along or near the "main highway" of the Amazon—the main body of the river along which Orellana and his tiny band made the first descent by Europeans. The span's combination of rich alluvial soils, excellent fishing, and easy access to riverine transport has attracted people since pre-Columbian times. Until a few decades ago, the várzea was inhabited and managed by smallholder farmers and fishermen, who tended to practice biologically diverse subsistence production based on the original indigenous methods. The advent of large-scale agriculture and commercial fishing operations is having a deleterious effect on both these cultures and these ecosystems. The current economic tumult and anti-environmental government orientation in Brazil only adds to the uncertainty about the future.

What are savannas, and how are they created?

The term "Amazonia" conjures up visions of verdant rainforests, but a significant portion consists of savannas. In the tropics, savannas are grasslands that run the gamut from extensive grassland with few trees to a dry forest with mostly trees and little grass. Amazonian savannas usually harbor at least some trees and typically feature a pronounced dry season, usually lasting longer than 3 months. This flora must be able to endure not only drought but also searingly high temperatures in tropical sunlight unprotected by a forest canopy. The predominant trees in these Amazonian savannas—like the sandpaper tree (*Curatella americana*)—typically have thick bark and leathery leaves adapted to survive both the high temperatures and the fires that often sweep through.

Indigenous people are known to periodically burn savanna to increase the number of young plants appealing to game species, to facilitate hunting of game species, and to expedite travel through the grasslands. Some scientists have postulated that all tropical savannas were in fact created by this form of burning, but research has revealed that savannas date from long before the first peopling of the Americas. Current thinking is that both soil and climate play a critical role in savanna formation and that fires—whether set by people or sparked by lightning strikes—maintain and perhaps even expand savannas.

Amazonia is home to many sizeable savannas like the Rupununi in southwestern Guyana or the Paru that straddles the Brazil–Suriname border. One of the largest is the famous *llanos*, the massive floodplain of the upper Orinoco River that stretches into both Colombia and Venezuela. The llanos is a seasonally flooded grassland with both forest islands and riparian (riverbank) forest that extends more than 30,000 miles (48,300 km) and can be covered with 3 feet (1 m) of water during the rainy season.

While ecotourists often bemoan the difficulty of observing rainforest wildlife, the llanos offer astonishing concentrations of wildlife, particularly in the dry season. Flocks of 70 species of waterfowl like egrets, herons, ibises, spoonbills, and storks (including the magnificent Jabiru, *Jabiru mycteria*, which stands almost 5 feet [1.5 m] tall) are found in astonishing numbers, and one can also see capybara (*Hydrochoerus hydrochaeris*), jaguar (*Panthera onca*), giant river turtles (*Podocnemis expansa*), and the fearsome and highly endangered Orinoco crocodile (*Crocodylus intermedius*). Much of the llanos has been given over to cattle, as ranchers found this savanna ecosystem much more suitable for cattle ranching than is the rainforest. The explosion of interest in ecotourism has seen cattle ranches augmenting income by creating wildlife viewing opportunities. This practice was common in Venezuela, but the collapse of the Venezuelan economy and the cessation of hostilities in Colombia means that this sustainable aspect of economic growth will more likely thrive in the Colombian llanos.

Another enormous Neotropical savanna is the cerrado of Brazil that forms the southeastern border of Amazonia. The soils are deep, sandy, and nutrient poor. However, the addition of massive quantities of fertilizers can make these soils highly productive in terms of intensive agriculture. As in the case of the llanos in Colombia and Venezuela, ranchers found this ecosystem easier to farm than rainforest and the creation of massive farms of soy monoculture over the past three decades has replaced enormous swaths of cerrado savannas—more than half the original cerrado has been converted to cropland over the course of the past few decades. The "Cerrado Pledge"—a voluntary commitment by large companies to halt deforestation resulting from intensive agriculture—was issued in 2017. However, several major commodity firms that purchase soy from cerrado lands, and China (perhaps the world's major consumer of cerrado soy), have not signed the pledge. And some of the corporations—like Cargill—that signed the pledge appear to not be meeting their targets.

What are tepuis, and what makes them unique?

Older than the Andes, *tepuis* are tabletop mountains—also known as mesas—that stretch across much of northern Amazonia, from Colombia east to Suriname. They are the remnants of an ancient sandstone formation that once covered the region, the majority of which has weathered away. Most of the tepuis,[5] which number well over 100, are found in Venezuela in a region known as the Gran Sabana ("The Great Savanna"), much of which is found in Canaima National Park. The highest tepui, Pico da Neblina ("Clouded Peak"), lies on the Brazilian side of the border and reaches 9,827 feet (2,995 m); it was declared a national park in 1979. The world's tallest waterfall—Angel Falls, at more than 3,200 feet (975 m)—plummets from the Auyan Tepui in Venezuela. So great is the drop that much of the water dissipates into mist on the way down.

Unlike connected mountain ranges, tepuis stand alone, and this isolation combined with their vertical walls and flat summits presents a breathtaking and unworldly air. Ethnobotanist Richard Schultes referred to them as "sentinels of a mysterious past."

The word "tepui" comes from the language of the Pemon people who inhabit the Gran Sabana and is said to mean "House of the Gods." To the west in Colombia, the Chiribiquete tepuis (also in a national park) were similarly venerated by tribal peoples who expressed their reverence by decorating caves at the bases of these mountains with paintings of shamanic art.

Rising out of the Amazon rainforest and savanna, these mesas inspire all who visit them, not just the indigenous inhabitants. Bordered on most sides by massive sheer cliffs and standing astride the point where Brazil, Guyana, and Venezuela all meet, Mount Roraima is one of the most stunning of the tepuis. Although supposedly first sighted by Sir Walter Raleigh and his compatriots in 1595, Roraima was not climbed by Europeans until 1884, when the British colonial administrator Everard im Thurn discovered (or, more likely, his Amerindian guides showed him) a trail to the summit. Im Thurn recorded his findings in memorable terms:[6]

> For all around were rocks and pinnacles of rocks of seemingly impossibly fantastic forms, standing in apparently

> impossibly fantastic ways—nay, placed one on or next to the other in positions seeming to defy every law of gravity—rocks in groups, rocks standing singly, rocks in terraces, rocks as columns, rocks as walls and rocks as pyramids, rocks ridiculous at every point with countless apparent caricatures of the faces and forms of men and animals, apparent caricatures of umbrellas, tortoises, churches, cannons, and of innumerable other most incongruous and unexpected objects.

Im Thurn's descriptions of Roraima were said to be the inspiration for Sir Arthur Conan Doyle's 1912 novel *The Lost World,* in which British explorers summit a Venezuelan tepui and encounter a plateau teeming with dinosaurs. Indeed, these Amazonian massifs are at least partly responsible for inspiring an entire genre of fiction devoted to tropical lost worlds filled with extraordinary creatures, from *King Kong* to *Jurassic Park.*

While no pterodactyls or stegosauruses live atop these stunning mountains, unique flora and fauna are plentiful. The tepuis harbor many endemic insect-eating plants, for example, an adaptation that allows these species to derive nutrients that they cannot extract from the poor soils atop the massifs. One of the most peculiar is the tiny—usually less than an inch (1 cm) long—pebble toad, known to exist only on the tops of two Venezuelan tepuis. Like an amphibian armadillo, the pebble toad (*Oreophrynella nigra*) rolls into a ball when threatened. Unlike an armadillo, however, this little frog will roll down a slope to escape a predator.

Because of the steep escarpments that form the sides of the tepuis, many of the species atop these mountains have evolved in isolation. Furthermore, climatic conditions differ strikingly from the surrounding rainforest. Because the canopy absorbs most of the light striking the rainforest, little sunlight reaches the forest floor. Temperature variation within the forest is minimal. Conditions are almost always wet and humid. Atop the tepuis, however, plants and animals must adapt to ferocious heat during the day and chilly temperatures at night. Unmitigated by a canopy, torrential rains pour down directly onto them. Moreover, there exists little vegetation to absorb the water, which then gushes off the tepuis in stunning waterfalls, such as Angel Falls, leaving drought-like conditions on the summit. The resulting ecosystems atop these tepuis, 3,300 to 9,800 feet (1,000–3,000 m) above the Amazon rainforest canopy, are entirely unique. They are truly islands in the sky.

3

RIVERS

The Amazon is a river of unparalleled vastness. Born high in the southern Andes of Peru, the Amazon flows north and then east, forming a national boundary with Colombia before crossing much of northern Brazil and then emptying into the Atlantic, draining an area of about 2.7 million square miles (7 million km^2). Headwaters that originate in Bolivia, Colombia, and Ecuador also flow into the Amazon as it makes its way to the sea.

The Amazon drains the world's largest rainforest and the most immense river basin, which harbors the most extensive assemblage of flooded wetlands known. Its basin represents the biggest drainage system on the planet and is more than twice the size of the Congo Basin. The Amazon drains almost 40% of the South American continent, with more than 1,000 rivers emptying into it as it makes its way from the Andes to the Atlantic.

The Amazon is the longest river on the planet and so broad in its lower reaches that in Brazil it has been called *O Rio Mar*: the "river ocean." In the rainy season, its mouth is about 300 miles (482 km) wide—greater than the distance from New York to Washington—and more than 25 times wider than any point on the mighty Mississippi. So long and deep is the Amazon that oceangoing vessels can sail from the Atlantic Ocean into the

Amazon Basin and then as far west as Iquitos, Peru—fewer than 200 miles (320 km) from the Pacific.[1]

By many multiples, it produces the greatest annual discharge of any river: as much as 17% of all freshwater released into the oceans, some 10 times more than the Mississippi, and over 50 times more than the Nile. Biologist Michael Goulding estimates that it discharges 57 million gallons (215 L) per second, such that, in 2 hours, the river could supply the annual water needs of New York City. So much freshwater gushes from its mouth that it reduces ocean salinity more than 100 miles (160 km) into the Atlantic. And daily, more than 100 million cubic feet (2.8 million m^3) of sediment empty from the Amazon into the Atlantic. Geographer John Hemming calculates that the Amazon exceeds the flow of the next eight largest rivers combined.

The mouth contains one of the world's largest riverine islands: Marajo, which is only slightly smaller than Switzerland. The Amazon's mouth also is home to the *pororoca*, at 12 feet (3.6 m) one of the highest tidal bores known. The *pororoca* has become popular with surfers who demonstrate their fearlessness not only by riding the great waves but also by dodging enormous tree trunks and errant black caimans that sometimes appear.

At times, the great river's tremendous flow is augmented: from November to June, heavy seasonal rains in the Peruvian and Ecuadorian Andes generate annual flooding along the Amazon. The river tends to reach its lowest point in August and September and its highest point in April and May.

Its two major source rivers—the Marañón and the Ucayali—originate in the Peruvian Andes and flow north where they join just south of the port of Iquitos to form the main river. This waterway turns east, gathering volume as it forms the southeastern border of Colombia. At the southernmost corner of Colombia, the river flows south of the major Amazon port city of Leticia and enters Brazil, where it is known to the Brazilians as the Solimões until its confluence with the Rio Negro, which enters from the north at the city of Manaus. From Manaus east until it enters the Atlantic Ocean, the river is called the Amazon (though, outside of Brazil, it is known as the Amazon from Iquitos, Peru, forward).

Numerous tributaries provide another 10,000 miles (16,000 km) of navigable rivers—and 17 of these branch rivers stretch for more than 1,000 miles (1,600 km). The Rio Negro in northern Brazil is the second largest tributary in the world. In fact, a dozen Amazon branches are each larger than any European river. One of these, the Madeira in southwestern Amazonia, has a basin whose area exceeds that of every South American country except Argentina and of every country in Europe except Russia. Comprising 20% of the Amazon Basin, the Madeira Valley is about three times the size of Spain.

Where does the Amazon River begin?

As early as 1707, the redoubtable Jesuit cartographer Father Samuel Fritz hypothesized that the Amazon had its origin in the frigid highlands of the Peruvian Andes, in Lake Lauricocha in the department of Huanuco. The search for—and the argument over—the river's ostensible source continues to the present day.

Part of the disagreement stems from differing definitions among cartographers of what precisely constitutes a river's source. One school of thought dictates that a river's origin is the most distant point from the river's mouth in which a drop of water enters the system—in other words, the farthest point upstream. The other definition is the most distant point on the stream that conveys the most water into the river. Fritz believed that the Marañón River was the ultimate source of the Amazon because it carried the largest volume of water into the upper Amazon near Iquitos. Others posited the Ucayali River as the source because it seemed to transport water from a more remote location than did the Marañón.

Certainly, the most colorful attempt to solve this riddle was a 1969 expedition led by Captain Loren McIntyre, a US Navy World War II veteran who had been seconded to the American Embassy in Lima after the war to teach gunnery tactics to the Peruvian military. A gifted writer and autodidact, McIntyre taught himself Spanish, Portuguese, and photography, and he soon became *National Geographic* magazine's preeminent photojournalist in Latin America during the latter half of the 20th century.

Prior to his expedition, McIntyre sorted through historical sources as well as newspaper articles and reports claiming to pinpoint the Amazon's origin at different places in the Peruvian Andes. Local geographer Carlos Penaherrera del Aguila had pinpointed the towering snow-capped mountain known as Nevado Mismi—more than 18,000 feet (5,500 m) tall and about 100 miles (160 km) west of Lake Titicaca—as the ultimate source.

With the sponsorship of the National Geographic Society, McIntyre and two colleagues made the excruciating trek up to and then up the mountain, where he observed that the Carhuasanta Creek flows off Mount Mismi and eventually enters the Rio Apurimac, which ultimately empties into the Ucayali. The pond below the Carhuasanta Creek was proclaimed the true source of the Amazon and named Laguna McIntyre in the explorer's honor. The expedition is memorably detailed in Popescu's 1991 book, *Amazon Beaming*.[2]

In 2014, explorer John Contos was kayaking in several of the most remote waterways of the Peruvian Andes when he hypothesized that the actual headwaters of the Amazon were not the Carhuasanta Creek on Nevado Misti, but rather the Mantaro River coming off the Cordillera Rumi Cruz to the west. Employing high-resolution satellite imagery and detailed topographic maps to buttress his conjecture, Contos then hiked and kayaked the relevant portions of both the Apurimac and the Mantaro, concluding that the Mantaro was 48 miles (77 km) longer.

Some challenge Contos's conclusions because the Mantaro River dries out for about 5 months of the year, though much of this is caused by the Tablachaca dam, which was built on the river in 1974. If Contos's conclusions are accepted, however, the additional length means that the Amazon stretches farther than the Nile and is therefore the longest river in the world. Meanwhile, the debate continues.

What are the various river types in Amazonia?

Alfred Russel Wallace (1823–1913) can be considered the most important scientific explorer in the history of the Amazon. Ironically, in the annals of science, he is best known for 8 years of research in the rainforests of the Malay Archipelago, twice as long as his 1848–1852 sojourn in Amazonia. It was in the Maluku Islands—famously, during a malaria-induced dream in a hammock—that he devised a theory of evolution that he then sent to Charles Darwin, who had already arrived at a similar conclusion.[3,4]

Wallace traveled the length and breadth of much of Brazilian Amazonia, as far north as Venezuela and as far west as Colombia. His travels and studies resulted in extensive collections, observations, and important publications on topics as diverse as palms, birds, pre-Columbian rock art, and tribal cultures and languages. And he also provided some of its first descriptions of the three main river types in Amazonia: black, white, and clear.

Wallace conducted much of his Brazilian research along the enormous and aptly named Rio Negro, which is not only the Amazon's biggest blackwater river, but also the sixth largest river in the world (and second largest in terms of discharge). At first sight, he wrote, "We might have fancied ourselves on the river Styx, for it was black as ink in every direction." Yet, on closer inspection, he realized the water itself was not actually black but "a pale brown colour."

Most of these so-called blackwater rivers course through northern Amazonia, draining the rainforests on nutrient-poor sandy soils. As a result, these blackwater rivers rank as some of the world's most nutrient-poor rivers, as devoid of nutrients as rainwater in some cases. Ancient Precambrian rocks—among the world's oldest—have been eroded for millions of years, meaning much of the nutrient content has long been weathered away.

Blackwater rivers represent a freshwater tea created when plant matter that has fallen or washed into the river fails to completely decompose. The sandy soils surrounding blackwater rivers support few microorganisms that would otherwise decompose the leaf litter remnants, and the high concentrations of tannins and other plant compounds remain suspended in these rivers, giving them their characteristic color and making them highly acidic.

Early explorers referred to blackwaters as "rivers of hunger" due to their paucity of the large game fish common in other Amazonian rivers. Ironically, though, the lower Rio Negro harbors the most diverse fish fauna of any freshwater river in the world, having twice as many species (700) as all the rivers of western Europe combined. Most of these species are relatively small in size.

Though the Brazilian Rio Negro is the best-known blackwater river, there are major blackwater rivers found elsewhere in northern South America. The Caroni and the Caura in northern Amazonia empty into the Orinoco in Venezuela, while the blackwater Apaporis and Vaupes are two major rivers in the Colombian Amazon and serve as home to many relatively traditional indigenous tribes.

Amazonian whitewater rivers—despite the name—are more of a café-au-lait muddy brown than white. The color is due to sediments derived from the Andes, where many of these major Amazon tributaries, like the Madeira in Brazil, the Caquetá in Colombia, the Napo in Ecuador, and the Marañón in Peru, originate. These muddy rivers begin as clearwater streams in the eastern Andes, where they start their descent into the Amazonian lowlands, gathering sediment along the way. As such, in the words of Michael Goulding, "the Andes are the main nutrient bank for the Amazon." And the main body of the mighty Amazon—called the Rio Solimões in Brazil—is a whitewater river.

Unlike blackwater rivers, there is considerable rise and fall to whitewater rivers and—at maximum flood stage—they deposit rich sediments along the banks of the *várzea* swamp forest, not unlike how the Nile once deposited fertile soils along its banks, giving rise to Egyptian civilization. It is these rich soils in Amazonia—far more productive than the nutrient-poor ones in the interfluvial terra firme forests—that made possible the enormous population densities of Amerindian peoples who lived along the banks of the Solimoes in pre-Columbian times.

Several whitewater rivers—like the Juruá and the Purus in southwestern Brazil—are not born in the Andes but transport sediments that originated in their own headwater floodplains. Whitewater river ecosystems usually harbor vast numbers of mosquitoes and other flying insects, which is why many tourist lodges are constructed near the relatively bug-free blackwater rivers whose nutrient-poor waters—which in turn lack the bacteria, protozoans, phytoplankton, and zooplankton on which blackflies and mosquitoes feed—are shunned by these hungry insects.

The third category is clearwater—originally called *blue water rivers* by Wallace. The major clearwater rivers drain the Brazilian Shield in the southeast Amazon, flowing over crystalline rocks and emptying at their northernmost point into the Amazon. As is the case with blackwater rivers, the ancient headwater soils are heavily weathered and release little sediment and few nutrients into the ecosystem. Because of their clear water, rivers like the Tapajos, the Tocantins, and the Xingu rank among the most beautiful bodies of water in South America.

What are some of the characteristic aquatic habitats in Amazonia?

Covering such an enormous area, Amazonia contains a wide variety of aquatic habitats. Some of the most characteristic are *oxbow lakes*, named for their distinctive horseshoe shape. Oxbows are formed when a bend in a meandering river is isolated as a result of a river changing course. These lakes range from 5 to 15 feet (1.5–4.5 m) in depth.

Oxbow lakes are most common in western Amazonia. Wildlife thrive in these calm lagoons whose banks attract increasing numbers of ecolodges. Oxbows often harbor thriving populations of endangered species like black caiman (*Melanosuchus niger*) and giant otters (*Pteronura brasiliensis*) that have been extirpated in many other locales. Giant Victoria water lilies (*Victoria amazonica*) also flourish here, as do the bizarre, pterodactyl-like hoatzin birds (*Opisthocomus hoazin*).

Floating meadows, or mats, represent another unusual Amazonian ecosystem. Typically found in nutrient-rich whitewater rivers and consisting primarily of species like paspalum grass (*Paspalum repens*) and the well-known water hyacinth (*Eichhornia crassipes*)—and sometimes colonized by vines, shrubs, and even small trees—these mats can exceed a square mile (2.5 km^2), reaching a size at which they occasionally pose navigational hazards. Exposed to direct equatorial sun most of the day and thriving on the rich nutrient load characteristic of whitewater rivers, floating meadows represent an important and productive ecosystem within an ecosystem.

These floating islands serve as important nursery habitats for many fish species as well as for frogs, caimans, snails, crabs, shrimp, arachnids, and innumerable insects. Amazonian species like the electric eel (*Electrophorus* spp.), the wattled jacana (*Jacana jacana*), the manatee (*Trichecus inunguis*; at more than 350 lbs [158 kg], one of the heftiest known freshwater herbivorous mammals), and the capybara (*Hydrochoerus hydrochaeris*; weighing more than 200 lbs

[90 kg], the world's largest rodent) also thrive in and around these aquatic meadows.

Amazonia features several enormous river islands. Marajo at the mouth of the Amazon is the world's largest river island: bigger than Belgium, it is four times the size of Jamaica. Located astride the equator, it was once home to the advanced Marajoara indigenous culture, whose artifacts were excavated and studied by archaeologist Anna Roosevelt and her Brazilian colleagues. They estimated that this pre-Columbian indigenous population exceeded 100,000, making it one of the most heavily populated sites in ancient Amazonia. Today, the most numerous inhabitants of Marajo are the water buffalo (*Bubalus bubalis*)—with almost half a million of these bovines, the human population is outnumbered by about 30%. Western Marajo was originally covered by forest, although much of that has been logged. The eastern half is primarily savanna. A significant proportion of the island is flooded for several months of the year.

Two other sizeable river islands that rank among the world's largest are the Ilha do Bananal (slightly smaller than New Jersey) in the Araguaia River and Tupinambarana (larger than Connecticut). These river islands tend to have started as sediment deposits, which were then stabilized by colonizing grasses. This herbaceous vegetation in turn was then succeeded by woody vegetation, which then formed forests. Today, the larger river islands in Amazonia harbor resident human populations.

What is the Casiquiare Canal?

Once the European exploration and occupation of Amazonia was under way, these outsiders were soon traveling into some of the most remote corners of the great rainforest. Among the most intrepid voyagers were Spanish clergy searching for indigenous souls to "save"—and Portuguese soldiers hunting for indigenous bodies to enslave. One memorable encounter between these two groups occurred in what is now southern Venezuela in 1744. Padre Manuel Roman, a Spanish Jesuit, was headed south, ascending the Orinoco River, when he met a party of Portuguese slavers who—to his surprise—had traveled north from Brazil by boat. He personally confirmed that the Orinoco and the Amazon were connected when he voyaged south with them down the 200-mile Casiquiare Canal into Brazil.[5]

Based on the results of Father Roman's expedition, the renowned French explorer Charles Marie de La Condamine soon reported on the connection of these two great river systems to the French Academy of Sciences. His claim generated a debate among other scientists, many of whom declined to believe his assertion that two great separate river systems—the Amazon/Negro and the Orinoco—were in fact connected. The French geographer Philippe Buache, who pioneered the use of contour lines in cartography but who—unlike La Condamine—had never been to the Amazon, declared that "the long-supposed communication between the Orinoco and the Amazon is a monstrous error of geography." Buache went so far as to create a map of the area, adding a nonexistent chain of mountains between the Rio Negro and the Orinoco to support his bogus contention.

In fact, a Spanish Boundary Commission confirmed the existence of the canal as early as 1756. But the first detailed scientific study of the Casiquiare was carried out by Alexander von Humboldt and Aimé Bonpland in 1800. Traversing the entire length of the canal, they collected biological specimens and surveyed the landscape. At least partially in homage to von Humboldt, visiting the Casiquiare became something of a rite of passage for European biologists visiting the Amazon in the 19th century, as the Austrian Johan Natterer, the German Robert Schomburgk, and the British scientists Alfred Russel Wallace and Richard Spruce all made the arduous and bug-ridden trek.

The first major 20th-century scientific expedition to the canal took place in 1919, when the Harvard physician-geographer Alexander Hamilton Rice compiled detailed maps of the Rio Negro, the Casiquiare, and the upper Orinoco employing the most modern surveying instruments then available. In 1968, explorer Robin Hanbury-Tenison led a hovercraft expedition from Trinidad to Manaus that passed through the canal as well as over and through the fearsome Atures and Maipures rapids on the Orinoco, features that had long served as effective barriers to most of the outside world.

Though universally referred to as a "canal," the Casiquiare is actually a river that splits off from the upper Orinoco (in technical terms, it is a bifurcation of the main channel) in Venezuela and then travels almost southwest into Brazil where it becomes one of the two headwaters of the Rio Negro (the other being the Rio Guainia, which originates in Colombia). That the Casiquiare serves as the world's largest natural waterway linking two titanic (but otherwise separate) river basins adds to its sense of mystery and wonder.

Is there a coral reef in the Amazon?

Because of their extraordinarily high levels of biodiversity, coral reefs have been called "the rainforests of the sea." Covering less than 1% of the earth's surface, they are believed to harbor as many as 25% of all marine species, brimming with fish, mollusks, crustaceans, echinoderms, and sponges. These reefs are typically found in clear, warm, sunny, shallow, nutrient-poor waters.

In 1977, researchers dredging for specimens in the muddy waters at the mouth of the Amazon found fish and sponges characteristic of coral reefs in their net. They published their results, which at the time elicited little in the way of interest or follow-up. After all, conditions at the Amazon mouth were the opposite of those in which corals were known to thrive: the water was cloudy, cold, deep, and rich in the nutrients that the Amazon was carrying to the sea. In 2016, however, an international team of researchers led by Brazilians from the Universidade Federal do Pernambuco and the Universidade Federal do Rio de Janeiro returned to the area and made an astonishing find: a 600-mile-long coral reef that stretched from the waters off southernmost French Guiana southeast to the coast of the Brazilian state of Maranhão.

This coral reef is unlike any other known: it is relatively low in coral cover but very high in sponge and rhodolith cover.[6] More anomalous is that this reef teems with chemosynthetic microbes that can create nutrients from the chemicals around them rather than photosynthetically from sunlight, which is in limited supply. In this they are like the vestimentiferans—the deep-water tubeworms that thrive near hydrothermal vents near the ocean floor. Thus, the microbiology and geochemistry of this Amazon coral reef are unique.

Ongoing commercial activities, however, threaten this extraordinary discovery. In the early 2000s, petroleum engineers began finding evidence of huge oil fields off the coast of South America. The initial strike—the Libra field, 140 miles (225 km) from Rio de Janeiro—began producing in 2017. Exxon Mobil recently claimed to have located a titanic deposit off the coast of Guyana. And a consortium of major oil companies exploring at the mouth of the Amazon believe they have located 14 billion barrels of petroleum there—more than the proven reserves of Mexico.

The drilling presents more than a proximate threat to the reefs. The recent finds are in exceedingly deep water—more than 6,000 feet (1,800 m). Oil extraction at these great depths entails greater risks as leaks and spills are harder to pinpoint and plug. The Deepwater Horizon spill in the Gulf of Mexico—the worst in US history—was one such example, being at a similar depth to the new Amazon find. A spill even a fraction of the size of Deepwater Horizon could not only destroy the

Brazilian reef, but could also decimate the rich fishing grounds along the mangroves that cover the continental coast parallel to the reef. At the same time, the Brazilian government has made sharp cuts to government agencies responsible for environmental protection.

4

AMAZONIAN BIOTA

Plants

What is ayahuasca?

From Argentina to Australia, from Israel to Istanbul, a once-obscure Amazonian liana, admixed with a few other plants, is now celebrated—and even venerated—as a plant of power, knowledge, and healing that has spawned two state-recognized neo-religions.

Ayahuasca is—first and foremost—a liana native to the northwest Amazon; its scientific name is *Banisteriopsis caapi*. Another common name is *yagé* (pronounced "yah-HAY"). These two appellations predominate in Colombia, Ecuador, and Peru; the vine is also known as *caapi* among the tribes of northwestern Brazil and is more commonly called *hoasca* in the rest of that country. Any of these names may also refer to the hallucinogenic potion created by brewing this liana with other plants.

The origins of ayahuasca use are impossible to determine. However, there exist abundant archaeological finds of anthropomorphic figures, snuff trays, snuff tubes, and even actual snuff residue that demonstrate hallucinogenic plant use in the western Amazon as far back as 2000 BCE.[1] Harvard ethnobotanist Richard Evans Schultes—who conducted seminal field research on ayahuasca in the mid-20th century—wrote, "The drink, employed for prophecy, divination, sorcery, and medical purposes, is so deeply rooted in native mythology and philosophy that there can be no doubt of its great age as part of aboriginal life."[2]

Ethnobotanist Constantino Torres, an authority on the history of ayahuasca and hallucinogenic snuffs, notes that the Jesuit missionary Jose Chantre y Herrera provided the earliest known account from the Marañón River region of the Peruvian Amazon in 1675: "The diviner hangs his [hammock] in the middle or takes himself a bench or a small platform and next to it places a hellish brew called [ayahuasca], remarkably effective in depriving of the senses. He makes a tea of the vine or bitter herbs, which after much boiling will become very thick. As it is so strong to disrupt judgement in small quantities."

The missionary's reaction to his experience is in keeping with the response of most historical ecclesiastic chroniclers when they encountered mind-altering plants and fungi employed by indigenous peoples of the New World: the clergy demonized and condemned these substances and mixtures, whether peyote in the north, magic mushrooms in Central America, or ayahuasca, yopo, and epena snuffs in Amazonia.

Equally characteristic is what transpired when the first ethnobotanist encountered ayahuasca: he drank it. In 1851, Richard Spruce stumbled upon an ayahuasca ceremony among the Tukanoan peoples on the Uaupés (Vaupés) River in northwestern Brazil, near the Colombian border. After the conclusion of the ritual, Spruce ventured into the surrounding forest to collect the vine in flower, necessary for making a precise scientific identification. Realizing that this represented a species unknown to science, Spruce named it *Banisteria caapi*,[3] thereby honoring and immortalizing the Tukanoan name for the vine: *caapi*.

Ayahuasca admixtures—plants added to the potion with the intention of altering the type, intensity, and duration of the experience—represent complex and fascinating aspects of the story. More than 100 different plants from 40 different families have been reported as additives to the brew; though most are flowering plants, at least one is a gymnosperm and another is a fern. The two most important admixtures, however, are either the shrub chacruna (*Psychotria viridis* of the coffee family, Rubiaceae) or the liana ocoyage (*Diplopterys cabrerana* of the Malpighiaceae, the same family as ayahuasca). *Psychotria* and *Diplopterys* contain hallucinogenic tryptamines (a type of alkaloid) that prove inert when consumed orally unless they are activated by the presence of compounds known as monoamine oxidase inhibitors. The ayahuasca (*B. caapi*) vine contains psychotropic alkaloids in this class, meaning the combinations of these plants produce strikingly more potent and profound effects than a potion prepared from either species. How shamans living in a rainforest comprising tens of thousands of plant species discovered the appropriate blend to induce otherworldly visions and insights remains a riddle.

In the Amazon, the brew is typically prepared by boiling the stem of the ayahuasca vine with the admixtures for several hours, producing a thick and highly bitter brew that is then consumed in small doses. Approximately 20 minutes after the initial dose, the subject usually experiences the onset of dizziness and nausea and then vomiting and/or diarrhea—purges that the shaman insists are the cleansing necessary to initiate the healing process. Within the next hour, visions commence, often inducing fear, stress, or even terror and frequently followed by scenes of unsurpassed loveliness and spiritual illumination. Participants in traditional ayahuasca sessions sometimes report the ability to communicate telepathically with the shaman guiding the ceremony—so much so that the first alkaloid isolated from the ayahuasca vine was named "tel epathine."[4,5]

Amazonian shamans imbibe ayahuasca to diagnose, treat, and cure illness and claim that the potion empowers them to see into the future, ward off misfortune, and provide protection against jealousy and negativity. Schultes painted a vivid portrait of the shamanic perspective on the potion:[6]

> [Ayahuasca] can free the soul from corporeal confinement, allow it to wander free and return to the body at will. The soul, thus untrammeled, liberates its owner from the realities of everyday life and introduces him to wondrous realms of what he considers reality and permits him to communicate with his ancestors. . . . Ayahuasca ("vine of the soul")—refers to this freeing of the spirit. The plants involved are truly plants of the gods, for their power is laid to supernatural forces residing in their tissues, and they were divine gifts to the earliest Indians on earth.

Given these exalted claims, it is unsurprising that those whom Western medicine has failed to cure venture into remote Amazonia in search of shamans and their healing plants. Sufferers of anxiety, chemical dependency, depression, infertility, insomnia, posttraumatic stress disorder (PTSD), sexual trauma, stress, and end-of-life anxiety travel to the rainforest in search of cures. Such is the demand that ayahuasca tourism lodges have sprung up throughout Amazonia, particularly around Iquitos, Peru. Sometimes things go well, and people are healed, healers are honored and fairly compensated, jobs are created, and the financial incentive for protecting the forest and the culture increases. However, this is an unregulated industry and sometimes subject to abuse and fraud.

Why are bromeliads known as the aerial aquaria of the Amazon?

Epiphytes are plants that grow on other plants (typically trees) but—unlike parasites—extract no nutrition from the host. Instead, they opportunistically gather nutrients from the rain, from dead leaves, and from animal remains and droppings. More than 15,000 species of epiphytes are known from Central and South America. Since so many flourish only in or near the forest canopy, undoubtedly numerous others have yet to be collected and classified. Tropical cloud forests tend to host the greatest diversity of epiphytes, but as many as one-quarter of the botanical species in lowland rainforests may consist of these types of plants.

More than 80 families of plants contain epiphytes, including philodendrons, cacti, ferns, and orchids. One family unique to the New World tropics and subtropics—the *Bromeliaceae*—harbors some of the most extraordinary adaptations and ecological partnerships known from the Amazon rainforest.[7]

There exist more than 3,400 species of bromeliads, but the two best known to the general public—the pineapple and Spanish moss—bear little resemblance to most members of the family. Perhaps the most emblematic species of this group in the rainforest are the remarkable tank bromeliads. These plants—usually found within or just below the canopy—feature tightly overlapping leaf bases that collect water into several small pools or a central tank.[8] These arboreal cisterns can contain over a half-gallon (2 L) of water, with one enormous terrestrial species from eastern Brazil accommodating more than 10 gallons (38 L).

Bromeliad tanks serve multiple, vital ecological purposes. They offer aquatic microhabitats high above the forest floor. The canopy layer has been called a "nutrient desert" because there is no soil layer from which epiphytes can draw sustenance. Consequently, bromeliads have evolved numerous intriguing adaptations that permit them to survive and thrive in the high frontier and ensure by their presence that the canopy is not a nutrient desert. The tanks themselves keep the plants watered throughout the year. And their characteristic sword-shaped leaves are covered with *trichomes*, highly modified hairs that efficiently absorb rainwater rich in organic compounds produced by tree and liana leaves growing above the bromeliads.

The tanks also catch rotting leaves, twigs, bark, and flowers, as well as animal droppings and carcasses, which are decomposed by aquatic bacteria and other microorganisms. This process releases nutrients absorbed by the plant itself and also serves as the basis for an astonishing ecosystem of aerial aquaria high above the forest floor. Large protozoans, tiny crustaceans (known as ostracods), and mosquito larvae consume the smaller microorganisms, and these in turn attract insects, which then serve as prey for bats, birds, frogs, and tiny salamanders. The small vertebrates are then eaten by larger birds, lizards, monkeys, and snakes. Other creatures documented to be a part of this arboreal rainforest ecosystem include worms, snails, and even crabs. All the while, the plant receives nutrients derived from the droppings and remains of all these creatures.

And, just as the tank serves as an aquarium, the older dead and decomposing leaves surrounding it form an arboreal terrarium. Especially in the dry season, this moist decaying vegetation offers a refuge for ants, centipedes, scorpions, spiders, frogs, lizards, and snakes.

In 1913, Costa Rican biologist Claudio Picado carried out the first survey of tank bromeliad fauna and counted 250 species. A more recent and oft-cited study carried out in the Ecuadorian Amazon by Peter Armbruster of Georgetown University[9] analyzed 209 bromeliads and found 11,219 creatures of more than 300 species. A complementary botanical study by Barbara Richardson estimated a hectare of lowland forest could contain 175,000 individual bromeliads holding 13,000 gallons (50,000 L) of water. In fact, some rainforest trees are so festooned with tank bromeliads that their branches have been known to break off under their own weight, especially after a heavy rain.

Studies of bromeliad–animal interactions continue to yield novel and astonishing findings. Many frogs lay their eggs in the bromeliad tanks—in fact, some species spend their entire life cycle within the plant; these animals are so tied to these angiosperms in this relationship that they are known as "bromeligenous frogs."

Other poison dart frogs deposit their eggs at ground level, where they are cared for by the male. Once the tadpoles have hatched, the female hoists them on her back and—one at a time—carries them up the tree, where she deposits them in a bromeliad tank pool. The female visits the tadpoles every few days, depositing several unfertilized eggs into the water to feed her maturing offspring as they turn into frogs. Considering that the adult frogs are only about an inch (2.5 cm) in length and the trees sometimes reach more than 100 feet (30 m) in height, one can only marvel at the intricate links and dependencies between different species in tropical rainforests.

What is the traditional use of coca in the northwest Amazon?

Coca—not to be confused with coconuts or with cacao, the source of chocolate—is typically a 3- to 6-foot (1–2 m) bush or small tree native to northwestern South America. The leaves are employed as a masticatory by indigenous peoples of the Andes from Bolivia north to Colombia and also by the tribes of the Colombian Sierra Nevada. The leaves of two species they chew, *Erythroxylum coca* and *E. novogranatense*, are rich in cocaine and therefore a powerful stimulant. Furthermore, coca chewing provides other benefits as well: hunger suppression, prevention of altitude sickness (employed regularly by miners working in the high Andes), and pain relief, and it also yields minerals, vitamins, and proteins.

Coca chewing has been documented as far back as 3,500 years in the Andes, although some researchers posit that the custom may have begun thousands of years earlier. The plant was brought to Europe in the 1500s and become popular there in the late 19th century. Sigmund Freud was an early proponent, promoting its use as a stimulant and a potential treatment for morphine addiction. Cocaine became a component of various tonics and patent medicines; Thomas Edison, Ulysses S. Grant, Henrik Ibsen, and Jules Verne were all said to be consumers of coca wine. One of the original cocaine products is still popular today: Coca-Cola. Though it no longer contains cocaine, a key flavoring ingredient in the soft drink is extracted from leaves grown commercially near Trujillo in Peru.

There exists an unusual and widespread means of coca consumption in the Colombian Amazon and neighboring Peru. The Indians cultivate a variety of the plant known as *E. coca* var. *ipadu*. Unusually, since most of their agriculture is carried out solely by women, the *ipadu* or *mambé* (as it is known colloquially) is cultivated by the men in gardens solely devoted to this plant. When the leaves are ready to be harvested—never less than a year after the shrub was planted and usually after 18 months—a group of men trek to the garden and fill baskets with the leaves amid much good cheer and consumption of previously prepared coca powder. Hauled back to the roundhouse in great handwoven baskets, the leaves are carefully spread out and toasted on a large flat clay or iron plate that is more often used to bake cassava bread. The dried leaves are then placed into a hollowed-out tree trunk—which serves as a mortar—and pulverized with a sizeable wooden club, which serves as a pestle. The rhythmic thumping of the coca being ground to a fine powder echoes through the maloca (tribal longhouse) for hours almost every night.

Meanwhile, other men burn leaves of the *Cecropia* tree, which will be added to the coca powder to provide the alkaline substance that facilitates the release of alkaloids. The Indians may also mix in the leaves or the leaf ash of several others plants to fortify the effects of the *ipadu* or to imbue a certain desired flavor. Which plants best amplify the strength or improve the flavor of the coca powder are the source of seemingly endless discussions as the indigenous people consume the coca powder in their roundhouses each evening.

The *ipadu* powder is then stored in a hollowed out-calabash or—more commonly these days—a plastic container with a tight-fitting lid. Throughout the day, the container is seldom outside the reach of its owner who, when he (or more seldom, she) feels the need, uses a spoon or—in the case of the Indians of the Caqueta drainage—the leg bone of a tapir (*Tapirus terrestris*) as a spatula to scoop about a teaspoon of the powder and place it between the cheek and gum. Unlike the coca leaf prized by Andean cultures, the *ipadu* quid is not chewed, but allowed to gradually dissolve and be swallowed, at which point the user takes another scoop. Among the most prodigious consumers, such as the Yukuna tribe, a person may consume over a pound of the powder daily.

Unlike purified cocaine, *ipadu* is not addictive and offers many positive attributes: slight and pleasant mood elevation while staving off hunger, thirst, and fatigue. Could coca leaves, coca tea, coca chewing gum, or even *ipadu* powder one day become an internationally safe and effective stimulant and diet drug?

To the Indians living a traditional lifestyle, coca is employed to facilitate conversation and bind the community together, to protect both the culture and the forest, to cure, and to give offerings to the nature spirits. For much of the past half-century however, because of its ready conversion to cocaine, coca has been much more of a curse than a blessing outside of its ritual context: violence, death, deforestation, pollution, and corruption have all flowed from the murderous cocaine trade. Perhaps some of the lessons learned from the increasingly widespread legalization of marijuana in many countries might one day help us pursue a similar path with coca in its native form. In the meantime, the traditional uses of coca by its traditional users should be celebrated and protected.

What is curare, and why is it important?

The Spanish conquistadors invaded the Americas with several insuperable advantages over the indigenous inhabitants: firearms, steel, horses, and—above all, even if not seen as such—infectious diseases, against which the Indians possessed little or no immunity. But the Native Americans did have and use a weapon against which the interlopers had little defense and no antidote: arrow poison.

The European raiders of Hispaniola and the northern coasts of South American were hit with arrows and blowdarts whose tips were smeared with the toxic latex of the infamous manchineel tree (*Hippomane mancinella*), which literally drove them mad with pain and killed them within days. Contributing to the lethal effect was the Spaniards' lack of hygiene (and antibiotics), which meant that a deep wound even without a poison could result in a fatal infection.

When the Spaniards pushed into the South American interior, they encountered another fatal form of arrow poison: curare. This plant mixture relaxes the muscles of the diaphragm: the victim can no longer draw in a breath and suffocates.

Curare has always served first and foremost as a hunting poison rather than a weapon of war. Amazonian Indians are famed for their marksmanship: I have watched hunters knock small birds out of the canopy more than 90 feet (27 m) above them using only blunt-tipped arrows because they only wanted a few feathers for a headdress; once these feathers were extracted, the unharmed but indignant bird was released. A question arises: If these hunters are so accurate with their arrows and blowdarts, why would they need to employ poisons? Furthermore, why poison an animal you intend to eat? The answers to these questions also explain why curare became so important in Western medicine.

For the indigenous peoples of the Amazon, the preferred arboreal game species are monkeys, and Amazonian monkeys have tails. The hungry hunter might hit a primate who then wraps its tail around a branch, keeping it from falling to the ground. Curare in the arrow or dart acts as a muscle relaxant such that the animal can neither breathe nor hold on to branches and plummets to the pursuer below. Crucially for the hunters, curare poisons only if it enters the bloodstream, not the human digestive tract, where it is broken down into harmless components.

Charles Marie de La Condamine provided the first important scientific account of curare from the Amazon. Eight years after his 1735 arrival in South America, he recorded the use of curare by different tribes of blowgun hunters in both Ecuador (with the Yameos) and Peru (with the Tikunas). In his later writings, La Condamine noted that curare recipes varied according to the tribe preparing them and that some curares consisted of many plants while others contained very few.

Alexander von Humboldt made important observations in 1800, when he and his colleague Aimé Bonpland witnessed the preparation of curare among tribes of the upper Orinoco River, making them the first trained biologists to observe the process. In 1812, British explorer Charles Waterton traveled barefoot and in the rainy season from coastal Georgetown through the rainforest to the Brazilian border—a trek of more than 400 miles (640 km)—in his quest to obtain the curare of the Makushi, reputed to be the Amazon's most lethal. Though neither a formally trained biologist or physician, Waterton possessed an exceedingly creative and inquisitive mind, and he proposed experimenting with curare as a treatment for both rabies and tetanus, believing the muscle-relaxing properties of this indigenous poison might make it a life-saving medication.

Waterton's tireless promotion of curare and actual experiments with the poison attracted the attention of the European medical establishment. In 1855, the French physiologist Claude Bernard used curare to elucidate precisely how muscles and nerves interacted, a major medical breakthrough. And experimental use of curare to treat the rare autoimmune disease myasthenia gravis, which causes extreme muscle weakness, led to further advances in our understanding of the neuromuscular system. Physicians then developed curare into a drug that could relax muscles during operations, which revolutionized abdominal surgery. Today, synthetic compounds based on (and inspired by) the original rainforest plant extracts have replaced the natural products.

Northeast Amazon curares were usually based on lianas of the strychnine family and used on arrow tips; northwest

Amazon curares were usually made from lianas of the moonseed family and employed on blowdarts. Numerous other Amazonian plants have been discovered in use that give arrows and blowdarts a deadly effectiveness. One tribe in Ecuador was making arrow poison from a rare tree in the cinnamon family; another in Venezuela manufactured theirs from an Amazonian nutmeg relative. That these plants were employed as hunting poison indicates they cause a physiological response in the body, meaning that—like the better-known curares—they may hold therapeutic potential. With the advent of Western technology, however, has come the shotgun, and therefore the bow, the arrow, the blowdarts, and the blowgun are disappearing faster than the rainforest itself. The loss of the knowledge of how to make and use these extraordinary weapons is accompanied by the loss of which plants once served as the basis of curares unknown to the outside world.

And curare manufacture entailed not only which plants to employ as ingredients but also how to prepare them. So-called curare admixtures usually consisted of chemically inert plants added to the toxic mix. Though sometimes dismissed as foolhardy since these admixtures were not in and of themselves poisonous, laboratory research has in one particular case proven the Indians' ethnobotanical sophistication. Curares made from *Strychnos* lianas almost always contain plants of the black pepper family (Piperaceae), which are not toxic. What scientists have learned, however, is that compounds in these supposedly chemically innocuous pepper vines increase the bioavailability of the blood, meaning that the blood of the victim absorbs the poison much more quickly and lethally. In a rainforest of tens of thousands of species of flowering plants, again how did Indians learn to combine these? And—perhaps more importantly—what other secrets are being lost as these forests are felled and this wisdom is forgotten?

Why are lianas so important, and yet so poorly understood?

To most people, lianas and palms represent the most emblematic botanical denizens of the rainforest. Palms, however, are a family of plants, whereas lianas—woody vines—consist of members of many different plant families, including an Amazonian palm. But ever since Johnny Weissmuller's Tarzan swung through the jungle (mostly in the Los Angeles Country Arboretum near Pasadena, rather than in Africa or the Amazon) on a woody vine, lianas have been considered the rainforest plant par excellence.

Despite the fact that many lianas thrive in the uppermost levels of the canopy, they start out as small shrubs on the forest floor and grow up toward the light. Because they climb up, through, and over trees, they do not have to expend energy on structural support and can focus instead on leaf production, rapid growth, and the production of unusual chemicals to deter predation by the myriad vertebrates and invertebrates that inhabit the canopy ecosystem.

The sheer variety and ingenuity of how plants climb has long fascinated biologists; even Charles Darwin published a book on the subject in 1865. Amid this variability, there are four major pathways utilized by Amazonian lianas: twining (winding their shoots around another plant), tendrilling (using specialized organs to clasp other plants), root climbing (clinging to supporting vegetation with sticky roots), and hook climbing (attaching to other plants with backward-pointing hooks or spines). Along with the structural advantages conferred by anchoring, lianas thus can better compete with their involuntary hosts for the precious sunlight striking the roof of the forest. Simultaneously, the root systems of these woody vines are aggressively competing underground with these same trees for the water and scarce nutrients of the forest soil, meaning that lianas are battling with these trees both on top of the forest and underneath.

Since the earliest scientific studies of Amazonian forests, botanists have usually chosen to simply bypass lianas in their research, hampered as they were by plants growing more than 100 feet (30 m) above their heads. Even when they were searching for lianas, the challenges were legion: Richard Schultes spent years searching for the guarana vine (*Paullinia guarana*) in flower. When he finally encountered his quarry in the Colombian Amazon, he and his indigenous colleagues had to fell seven trees to obtain the blossoms from a liana that was snaking its way through the top of the forest canopy.

Indigenous peoples in these forests, however, can often distinguish lianas by the appearance of the stem or the fragrance of the bark. Their sophistication at identifying these species is matched only by their astonishing abilities to make full use of them. In tropical South America, Amerindian peoples have employed lianas as food, arrow poisons, fish poisons, cordage, stimulants, hallucinogens, snuffs, dyes, euphorics, sources of potable water and other drinks, coca additives, and a soporific to stun bees so that honey could be extracted from their hives. In terms of international commerce, Amazonian lianas like ayahuasca (the healing hallucinogen), barbasco (*Lonchocarpus* spp., source of the biodegradable pesticide rotenone), cat's claw (an immunostimulant), curares (*Chondrodendron tomentosum*, anesthesia), and guarana (*Paullinia guarana*, a stimulant) have proved to be valuable commodities.

Scientists have recently reached some surprising—and somewhat alarming—conclusions while studying lianas in the western Amazon and in Panama. Beginning in 1980, these biologists documented a rapid and major increase in liana abundance, presumably due to climate change since these woody vines are known to react positively to increases in carbon dioxide levels. Given the ongoing battles between lianas and their tree hosts, certain arboreal species known to be the favored hosts for lianas will decline. The result could be a change in forest composition and ecosystem dynamics as the creatures that rely on particular tree species suffer a competitive disadvantage and no longer serve as effective pollinators or seed dispersers.

What are strangler figs, and why are they not considered to be lianas?

Strangler figs are the botanical body snatchers of the rainforest. Their seeds—typically bird- or primate-dispersed—usually germinate on tree branches high above the forest floor. These seedlings shoot tendrils downward along the tree bole. In the course of their descent, the tendrils meet and fuse together—a process known as *anastomosing*—creating a botanical mesh. They then burrow into the forest soil in search of nutrients and water. Once rooted, the strangler then grows upward as woody vines along the host tree, in search of sunlight, all the while enveloping the supporting tree in a cruel, opportunistic, and often fatal embrace. If the host tree perishes, the strangler fig forms a ghostly tree with a hollow core. Even more ghoulish, the decaying remnants of the host tree serve as nutrients for the strangler fig. With typical mordant humor, the common name for "strangler fig" in the lingua franca of Suriname is *abrasa*—"the hugger." In Spanish-speaking Latin America, stranglers are known by the more prosaic *matapalo* ("tree killer").

To the rainforest visitor, the most obvious manifestation of strangler figs is as woody vines. Unlike lianas (which are also woody vines), strangler figs begin life in tree branches well above the forest floor, whereas true lianas start out on the ground and then ascend a host tree. The botanical term for plants whose life cycle is similar to the stranglers is "hemiepiphyte."

Though stranglers are murderous, these plants and other related figs serve as essential life-giving components of the rainforest ecosystem. Well over 1,000 animal species worldwide feast on figs, and most of these in the Amazon are birds (particularly cotingas, doves, motmots, parrots, tanagers, and toucans) and mammals (including bats, coatis, deer, monkeys, opossums, peccaries, tapirs, and tayras). But in Amazonia reptiles like the red- and yellow-footed tortoises (*Geochelone* spp.) and, incredibly, more than 100 species of fish are known to consume fig fruits, as are many invertebrates such as ants, dung beetles, snails, and even hermit crabs. Figs tend to be high in carbohydrates and calcium.

What makes figs so essential to life in the Amazon (and elsewhere in the tropics) is that they produce fruits year-round, often in exceptionally large amounts. In the Peruvian Amazon, one scientist counted more than 100 squirrel monkeys simultaneously feeding on a single fig tree at the Cocha Cashu biological station. Because few other plant families in the Amazon yield fruit all year round, many biologists regard figs as "keystone species" in that they reliably feed a diverse fauna during months when other foodstuffs might be scarce or nonexistent.

Why are palms the most useful group of plants to the indigenous peoples of the Amazon?

Linnaeus himself hailed palms as "Principes—The Princes of the Plant Kingdom." A coconut palm on a sandy beach represents the single most emblematic symbol of a tropical paradise. Festoon a stand of trees with lianas and palms, and many would deem it a tropical rainforest. In the words of Richard Schultes: "A panorama does not seem to be tropical unless palms occupy a distinct and conspicuously visible part of the flora."[10]

Unlike most trees of the rainforest, almost all palms can be recognized as such on sight. Their aesthetic appeal is undeniable, but their extraordinary utility is almost always underestimated by people outside the rainforest.

Associated primarily with the equatorial lowlands, palms thrive elsewhere as well. The Andean wax palm (*Ceroxylon quindiuense*)—at more than 200 feet (60 m), the tallest monocot in the world and much remarked upon by Alexander von Humboldt—thrives at over 10,000 feet (3,000 m) above sea level in the northern Andes. In the scorching and almost barren deserts of the Middle East, palms are truly the tree of life. Desert dwellers in ancient Egypt were devouring doum palm (*Hyphaene thebaica*, not date palm, *Phoenix dactylifera*) fruits 18,000 years ago. Due to their graceful appearances and bountiful yields, palm trees have served as sacred symbols in Buddhism, Christianity, Hinduism, Islam, and Judaism. And transplanted Chinese windmill palms (*Trachycarpus fortunei*) reach as far north as the shores of Scotland and the foothills of the southern Alps, where they adorn the shores of Lake Como in Italy.

But it is in the tropics that palms reach their greatest abundance, fecundity, and variety. Of the 200 genera found worldwide, about one-third of these are native to the Americas. Of the 67 genera in the New World, 34 genera comprising about 150 species are found in the Amazon. Ironically, some consider the diversity of palms in Amazonia depauperate: tiny Costa Rica—substantially less than 1% of the size of Amazon rainforest—harbors 109 palm species. Ecuador—4% the size of Amazonia—has 120 species.

Palms are notoriously difficult to study and collect. Botanists gather plant specimens in the wild by folding leaves into and then drying them in a plant press, the dimensions of which are a mere 12 × 18 inches (30 × 45 cm). However, palms produce some of the largest leaves in the world: in Amazonia, the leaf of the buriti (*Mauritia flexuosa*) palm reaches a length that exceeds 10 feet (3 m), while the African Raffia palm (*Raphia regalis*) leaf extends more than 80 feet (24 m).

But what Amazonian palms lack in diversity, they make up for in abundance: some pure stands of buriti and acaí (*Euterpe oleracea*) palms extend for dozens of miles. An extraordinary recent study of the South American rainforest determined the 20 most common tree species in the Amazon—and seven of those were palms.[11]

Ethnobotanists and food historians have long noted that the three plant families that have proved fundamental to the well-being of the human species are the Gramineae (grasses, such as corn, rice, and wheat), the Leguminosae (legumes, such as beans, peas, and soybeans), and the Arecaceae (the palms). Yet grasses and legumes play relatively limited roles in the lives of rainforest denizens, meaning that the importance of palms is even greater in these equatorial regions.

First and foremost, palms provide foodstuffs in prodigious quantities: a single bunch harvested from one palm can yield more than 1,000 fruits. In fact, only manioc is more important as a food source in places like the Vaupés and Miriti Paraná Rivers in the Colombian Amazon—and palms also provide dietary components like minerals, oils, proteins, and vitamins that manioc lacks. And not only are palms dietary staples of indigenous and peasant forest dwellers, but they are also sold in local markets (like maripa fruits, *Attalea maripa*), national markets (buriti ice cream), and even international markets (the now seemingly ubiquitous acaí juice).

The nutritional importance of palms extends far beyond the edibility of their fruit. Amazonian peoples also consume palm hearts (actually a cylindrical bunch of leaf bases that serve as the growing bud) and high-quality oils from the fruit, concoct breads from the starch of the stem, and prepare three different types of wine—one from the fruit, one from the trunk sap, and another from the unopened flower clusters.

Palm fruits serve as dietary staples for Amazonian birds (particularly macaws), fish (so much so that some palm fruits are employed to bait hooks by peasant fisherman), and mammals. In fact, two of the most important game animals in the Amazon forest—the ubiquitous collared peccary (*Pecari tajacu*) and the fearsome white-lipped peccary (*Tayassu pecari*)—feed on 25 and 37 species of palm fruits, respectively.

The second most important utility of palms in the Amazon rainforest is as shelter. Whether temporary lean-tos to wait out a sudden rainstorm, overnight camps for a hunting trip, or titanic longhouses (*malocas*) built to last more than a decade, almost all forest dwellings in Amazonia built solely of local materials contain palms, many consist of mostly palms, and a few consist of only palms.

The widespread reliance on these leaves for thatching is based on their internal strength, durability, and flexibility due to the presence of vascular bundles, silica bodies, and fibers that make them both strong and pliable. Studies on coconut palm (*Cocos nucifera*) leaves have found that the tensile strength of the leaf midrib is comparable to that of annealed aluminum, and palm thatch roofs in tropical Asia are known to have endured for more than 50 years. Despite the advent of corrugated tin in most corners of the lowland tropics, palm thatch roofs are still widely employed because they do not require a financial outlay (e.g., the leaves are freely available in the forest); they are far cooler under the equatorial sun than metal roofs; they are far quieter during the equatorial rains; they allow cooking smoke to escape, thus enhancing air quality; and they are lighter in weight, thus requiring a lighter frame for support.

The most famous student of these palms was Alfred Russel Wallace, co-creator of the theory of evolution. After 4 years in the Amazon collecting biological specimens, Wallace set sail for London in July of 1852, but his ship caught fire in the mid-Atlantic and he was forced to abandon almost all of his specimens and belongings. So important to him were his drawings of the palms used by his beloved indigenous colleagues that these sketches were one of the few items he chose to save, and they later served as the basis for his classic book, *Palm Trees of the Amazon and Their Uses*. In perhaps the most famous description of the utility of these princes of the plant kingdom ever recorded, Wallace noted the use of 13 different palm species employed along the Rio Negro as food, drink, musical instruments, harpoons, blowguns, blowdarts, cloth, roofs, doors, flooring, bowstrings, combs, boxes, bread, and fishing line.

In addition to the purposes elucidated by Wallace, palm fruits are also used as livestock feed, while the leaves and fibers are made into hats, mats, shoes, baskets, cordage, brooms, fire fans, roofs, walls, spears, bows, arrows, looms, graters, kitchen utensils, fencing, souvenirs for tourists, and backpacks. And the huge leaves of one species are employed as sails atop small boats in the Orinoco.

But it would be a mistake to suggest that the utility of palms is mostly limited to pre-industrial cultures: new and promising uses continue to be investigated and developed. For example, in Brazil, the starch from the stem of the buriti palm has been successfully tested as a component of gluten-free cookies and other baked goods. And acaí fruit shows promise for use in modern medicine as a (tasty) alternative to the (ghastly) oral contrast agent employed in magnetic resonance imaging (MRI) studies of the gastrointestinal tract.

The buriti palm is one of the most majestic, most easily recognized, and most common trees in the South American rainforest. This plant is found throughout the Amazon, and it especially thrives in swampy areas of little use for traditional agriculture. Buriti provides the Indians with food, drink, shelter, and clothing. The fruit produces as much vitamin C as citrus and as much vitamin A as carrots and spinach. It also yields an edible palm heart; a sago-like starch to make bread; a cork-like material employed for a variety of purposes like fishing floats, shoe soles, and bottle caps; an edible oil; and an industrial-quality fiber made into twine, sacking, nets, and hammocks. And fallen buriti palms are sometimes left in the forest to attract snout beetles that lay their eggs in the trunk. These hatch as white palm grubs, valued as a delicacy by the Indians for their high fat content.

The tastiest and most nutritious palm in Amazonia is the peach palm, also known as the pejibaye or chontaduro. Grown widely in indigenous and peasant gardens, it may be the single most balanced dietary staple of any rainforest plant product, containing carbohydrates, minerals, oil, protein, and vitamins. The fruit is borne in bunches that weigh up to 30 pounds (14 kg) and consist of up to 500 fruits, with as many as 13 bunches at a time on a single tree and producing as many as two crops per year. Not only rich in carbohydrates, it offers twice the protein content of banana and can yield more carbohydrates and proteins per acre than corn. It is widely available and treasured throughout both Central and South America, eaten boiled, cooked into a tasty porridge, turned into ice cream, and even used in craft cocktails. The peach palm will become a global crop in the future.

Another palm denizen of Amazonia is the tagua nut (*Phytelephas aequatorialis*), also known as the vegetable ivory palm—in fact, the genus name *Phytelephas* means "plant elephant." The hard white endosperm of the fruit ripens into a substance seemingly indistinguishable from elephant ivory. Tagua already appears in the tourist trade, particularly in Ecuador: carved tagua is made into earrings, chess pieces, beads, buttons, jewelry, keychains, dagger handles, even mouthpieces for bagpipes. In the hands of skilled artisans, tagua jewelry and art is exquisite; it is often suggested that increasing the supply of tagua for such purposes could conceivably reduce at least some of the demand for elephant ivory, thereby reducing the pressure on elephant populations. Before hardening into tagua, the unripe endosperm is delectable, tasting a bit like a coconut gelatin. Tagua palms also yield large leaves useful for thatch and a strong and lightweight wood.

In terms of international commerce, the most important Amazonian palm remains the acaí, whose annual export value from Brazil alone is said to exceed $100 million. The whitish stem, purple fruits, and feathery green leaves form an unforgettable tableau, emblematic of the lower Amazon. It is a multistemmed palm, often found growing in clumps in swampy or floodplain areas or planted in or near Indian and peasant gardens. Many indigenous peoples consider this palm to be of inestimable value because it is exceptionally abundant, is easy to fell, lacks spines, and has strong yet pliable leaves. They employ the roots to treat fevers and weave the leaves into durable and lightweight backpacks. Other indigenous groups value the roots as a vermifuge and the leaves as thatch.

The stems are used in light construction and as firewood, the inflorescences as brooms and fertilizer. Acaí produces an edible palm heart, but it is most highly valued for its fruit.

Acaí represents one of the "newest" fruits introduced into international commerce even though the provenance is ancient by Amazonian standards: Anna Roosevelt found it in pre-Columbian sites on Marajo Island in the estuary. Today, the fruit is made into juice, ice cream, porridge, popsicles, smoothies, liqueur, beer, vodka, and a savory steak sauce served at high-end restaurants in Rio de Janeiro.

Acaí's major importance as a food source is for the peasant populations in lower Amazonia, for whom it serves as a staple. "Acaí na tigela"—acaí in a calabash—is sometimes consumed more than twice a day and serves as an important source of calories and edible oil. Acaí has proved to be more fattening than milk due to its high lipid content, although it contains less calcium and protein. Given increasing global demand, plantations of acaí are expanding. Like the peach palm, acaí may one day be planted in Africa and Asia as well.

Did an Amazonian water lily serve as the inspiration for steel frame architecture?

As the largest city in the Peruvian Amazon, Iquitos has become an ecotourism mecca. Most come to enjoy the luxuriant forest flora and fauna, but some visit to experience the so-called mystic tourism of the flourishing ayahuasca lodges nearby. While the city itself offers relatively little to the experienced traveler, one notable stop on the standard city tour is the Casa de Fierro—the "Iron House." A striking structure built of iron sheets shipped over from Europe, it sits on the southeast edge of the Plaza de Armas. The Casa de Fierro dates from the late 19th century, the beginning of the Amazon rubber boom, when Iquitos served as the headquarters for Julio Cesar Arana's Peruvian Rubber Company and was awash in both money and European influence.

Local tour guides now seldom fail to claim that the Casa de Fierro was designed by the eponymously immortal French architect Gustave Eiffel (possibly true) and often claim that his design was inspired by the leaf structure of the giant Amazon water lily (not true), which in turn served as the inspiration for steel frame architecture and—therefore—skyscrapers.

Yet the tale they tell contains a botanically appropriate kernel of truth.

The world's largest water lily, the *Victoria amazonica*, is a botanical beauty of stupendous proportions: its circular leaves can span more than 8 feet (2.5 m). The netted venation along the underside so effectively distributes weight that a single leaf can support up to 100 pounds (45 kg), and an oft-photographed image of the lily is with a child seated on it as proof of the plant's size and strength.

Western discovery of *V. amazonica* is often mistakenly ascribed to botanist Robert Schomburgk. However, in the words of botanical historian Richard Mabey: "It was a measure of Britain's dominance of European botany in the mid-nineteenth century that its scientific establishment was able to blank out four separate [and earlier] discoveries by French and German explorers of the plant that became the sensation of the Victorian era."

While searching for new and useful plants in the employ of the Spanish crown, the Czech scientist Thaddeus Haenke encountered giant lilies in southern Amazonia near the Rio Mamore in Bolivia in 1801.[12] Haenke made sketches of the plant—a common practice by biologists of the time—but he apparently failed to collect specimens or take detailed notes.

The next scientist who encountered a Victoria water lily was French botanist Aimé Bonpland, who had traveled with von Humboldt on his tropical American foray from 1799 to 1804 and who was living in northern Argentina in 1820 when he found a giant water lily near the town of Corrientes, well to the south of Amazonia. Though this lily was closely related to the *V. amazonica* chanced upon by the ill-fated Haenke, subsequent research revealed it was a different and slightly smaller species, *V. cruziana*.

The first important scientific collection of the giant lilies was made by the German biologist Eduard Poeppig in the Peruvian Amazon in early 1832. Poeppig not only preserved parts of the plants in alcohol, but he also published a detailed description of his *V. amazonica* specimens in 1832, meaning that, from the perspective of Western science, he was the discoverer of the species.

On New Year's Day in 1837, Schomburgk—exploring British Guiana under the sponsorship of the Royal Geographic Society in London—encountered giant lilies along the upper Berbice River in the eastern sector of the colony. Though Poeppig had published his description of the same plant just 5 years earlier, his discovery was disregarded as Schomburgk and the British establishment rushed to celebrate the find as a means of honoring Victoria, who had recently been crowned queen. Part of the challenge thereafter was to develop a method to induce this Amazonian native to flower in England, no easy feat in northern Europe before the advent of widespread central heating. Joseph Paxton, an innovative garden designer and head horticulturist for the sixth Duke of Devonshire, caused a sensation in plant-mad England when he managed to get the plant to flower in the greenhouse at the duke's estate at Chatsworth in 1849.

Paxton did not rest on these (Amazonian) laurels. A keen observer of plants, he noticed that the strength and weight-bearing abilities of the Victoria lilies were due to leaf ribs radiating from the center that connected with cross ribs. He began constructing greenhouses at Chatsworth that employed this design. When the British government organized a competition to design an innovative and cost-effective structure for the Great Exhibition Hall of 1851 in Hyde Park, Paxton submitted a plan for a cast iron and plate glass edifice based on the venation of Amazonian water lilies. His submission, the "Crystal Palace," was accepted and constructed. The hall's architectural influence is felt to this day, particularly in the spectacular tourist greenhouses that grace major botanical gardens.

Animals

What is the origin of Amazonian fauna?

For much of the past 66 million years—the Cenozoic Era—South America was an island inhabited by an extraordinary group of mammals. There were bizarre ungulates (hoofed mammals) such as the now-extinct astrapotheres, litopterns, and notoungulates, as well as the more familiar anteaters, armadillos, opossums, and sloths.[13] Missing were artiodactyls (even-toed ungulates, such as deer and peccaries), carnivores, perissodactyls (odd-toed ungulates, like tapirs), primates, and rodents. This period of "splendid isolation"—a term coined by an American paleontologist—permitted the evolution of many unique and often colossal creatures in South America, most of which have since gone extinct.

The Phorusrhacids of South America, for example, were ferocious, flightless birds that reached a height of 10 feet (3 m). Rapid runners with enormous beaks and sharp talons, these so-called *terror birds*—in the absence of carnivores—filled the role of apex predators. But not all of the largest birds of prehistoric South America were flightless: with a wingspan of 20 feet (6 m), the *Argentavis* was the largest bird ever to have lived. Reptiles also attained great size, including crocodilians like *Gryposuchus*, which could reach a length of 36 feet (11 m). Turtles were similarly titanic: with an 11-foot (3 m) shell, *Stupendemys* was the largest freshwater turtle of all time; it is believed to be related to the current-day side-neck *Podocnemis* turtles of Amazonia.

The South American fauna began to change significantly in the middle Eocene, approximately 40 million years ago. This period featured the arrival of *waif dispersers*: creatures that floated out of rivers in western Africa on natural rafts when the distance between Africa and South America was much less than it is today as the continents drafted apart, and when westward prevailing currents facilitated this migration. First to arrive were the caviomorph rodents, the ancestors of Amazonia's agoutis (*Dasyprocta* spp.), capybaras (*Hydrochoerus hydrochaeris*), and pacas (*Cuniculus paca*).[14] The ancestors of all Neotropical primates also arrived from Africa via similar means.[15] African turtles may have floated over to South America without rafts: some turtles can float with their head above water and can survive long periods without food or water. Recent genetic analysis has shown that the ubiquitous Amazonian rainforest red-footed (*Chelonoidis carbonarius*) and yellow-footed tortoises (*C. denticulatus*) of today—as well as the giant Galápagos tortoises (*Chelonoidis* spp.)—are most closely related to the African hingeback tortoises (*Kinixys* spp.).

Massive changes in the South American fauna since the arrival of these foreigners are primarily due to what has been called the Great American Interchange. Approximately 4 million years ago, volcanic action created the Isthmus of Panama, forming a terrestrial bridge between North and South America, and animals began to migrate in both directions.[16] Multiple groups of organisms that ventured south from North America flourished and diversified in their new home and quite likely outcompeted the previous residents, driving them to extinction. In fact, many of the signature species most identified with the Amazon—bushmasters (*Lachesis muta*), coatimundis (*Nasua nasua*), fer-de-lance snakes (*Bothrops atrox*), giant otters (*Pteronura brasiliensis*), jaguars (*Panthera onca*), peccaries (*Tayassu* spp.), and tapirs (*Tapirus terrestris*)—are descended from North American ancestors.

Scientists estimate that there currently exist between 1,300 and 1,500 species of birds in Amazonia; this constitutes more than 10% of the world's avifauna. More than 1,000 species of frogs are known from the Amazon, and the high humidity and constant rains mean that frogs can roam throughout the ecosystem, from the forest floor to the canopy, unlike in the temperate zone where they are confined to the terrestrial proximity of puddles, creeks, swamps, and rivers. More than 3,000 species of fish thrive in the rivers and creeks of Amazonia, and, as with the frogs, new species are still being discovered and described. This ecosystem is home to 450 species of reptiles and approximately 430 mammals, primarily bats and rodents. Invertebrates dominate the Amazon: insects like ants, bees, beetles, butterflies, termites, and wasps, as well as arachnids like spiders and scorpions, tend to be much more ubiquitous than larger animals—estimates of invertebrate diversity in the Amazon rainforest range from 1 million to more than 20 million species; one recent estimate was that ants make up 30% of the animal biomass in the Amazon.

By comparison, flowering plants in Amazonia are estimated to number about 40,000 species, of which 16,000 are trees.

Do vampire bats drink human blood?

Vampire bats (*Desmodus rotundus*) are the embodiment of the term "bloodthirsty." Though most rainforest bats consume fruits, floral nectar (mostly carbohydrate), pollen (mostly protein), insects, and other small vertebrates, vampire bats are *hematophagous*: they feed solely on blood.

These nocturnal creatures are exquisitely evolved to exploit this peculiar lifestyle: thermoreceptors in their face help them locate precisely where blood flows close to the surface of a victim's skin. Capable of flying extremely quietly, they prefer to land near sleeping prey and approach on foot. In fact, vampires are extremely nimble, using their hind legs and large thumbs as feet to walk, jump, run, hop, or climb onto prey. There, the little mammal employs razor-sharp, canine-like and forward-pointing incisors to slice the outer skin of a slumbering animal, patiently lapping (not sucking) the blood as it oozes from the open wound. In fact, so sharp are vampire bat teeth that museum researchers must be extremely careful not to cut themselves when handling the skulls.

A protein in the bat's saliva that acts as an anticoagulant facilitates the feeding process by ensuring an uninterrupted flow of blood. The compound has been isolated and identified: its technical name is desmoteplase, but it has been given the more memorable moniker of *draculin* and is being investigated as a potential treatment for stroke due to its ability to prevent blood from clotting.[17]

A blood diet imposes several restrictions on the bat. Vampires are small and highly mobile creatures and cannot hibernate while fasting for lengthy periods, as do bears in the temperate zone. In fact, vampire bats cannot survive more than 72 hours without a meal. These creatures—who usually inhabit dead trees or caves in flocks that can number in the hundreds—will often regurgitate blood and share it with their roost mates. And to reduce chances of going hungry, a bat feasting on a victim will sometimes consume fully one-half of his or her body weight at a single meal, meaning the mammal has to begin digesting the blood and excrete waste products prior to departure to be light enough to achieve flight.

In the Amazon, vampire bats originally preyed primarily on large forest mammals such as peccaries and tapirs. However, with the introduction of massive numbers of cattle and pigs into deforested areas, they have shifted their preferences to these larger, more sedentary, and more abundant creatures. The result has been a population explosion of vampires in some locales, resulting in more predation on humans. In the past, such attacks were neither very common nor very dangerous, but the advent of large numbers of cows and pigs has also resulted in an increased incidence of rabies. Due to increasing deforestation and additional numbers of cattle and swine, this deadly virus is expected to continue to spread.

How dangerous are Amazonian spiders?

In a terrifying sequence in the 1962 British spy film *Dr. No*, James Bond awakens in his bed to find a huge and hairy tarantula crawling up his naked chest. Ever since, many Westerners associate the tropics with these hirsute behemoths. In fact, the Amazon teems with poisonous spiders, but tarantulas present little threat to *Homo sapiens*.

Approximately 900 tarantula species live around the world, most in the Americas. Common in the rainforests of Amazonia, members of the Theraphosidae family also thrive in cloud forests, grasslands, mountains, and even deserts. Solely equatorial they are not: an ebony species found in the hills near Folsom Prison about 100 miles (160 km) northeast of San Francisco was named after musician Johnny Cash: *Aphonopelma johnnycashi*.

All tarantulas are poisonous and have fangs, but they tend to be relatively harmless to people. In Amazonia, they prey on bats, birds, lizards, mice, snakes, and large invertebrates. They are such prodigious consumers of cockroaches and other vermin that some indigenous peoples of Amazonia allow them to live and nest in the thatched roofs of their dwellings as an effective means of pest control.

Though posing no life-threatening danger to British secret agents or anyone else in the tropics, tarantulas do have a weapon beyond their bite: tiny, barbed bristles known as urticating hairs that can be flicked into the air in such numbers as to form a small cloud. For humans, they can cause irritation, a rash, or, in rare cases, eye damage because these hairs can pierce the cornea.

For all their fearsome reputation as rainforest predators, tarantulas cannot chew: the mouth consists of a small opening that can only suck. Before ingestion, larger prey are typically drenched with digestive juices to partially dissolve them prior to being slurped up by the hungry arachnid.

The world's largest spider is an Amazonian tarantula: the Goliath birdeater (*Theraphosa blondi*), which has a leg span of 12 inches (30 cm) and fangs that can extend more than 1.5 inches (4 cm). They tend to live in silk-lined subterranean burrows. Goliaths feed primarily on earthworms, frogs, and insects, but also visit bird nests, where they employ their enormous fangs to puncture the shells so they can sip the eggs. And their name is no hyperbole: on rare occasions, Goliaths catch and consume small birds.

The English naturalist and explorer Henry Walter Bates had a memorable encounter with a Goliath in the rainforest near the mouth of the Amazon in the mid-1800s:[18]

> Two small [finches] were entangled in the [web] . . . one of them was quite dead, the other lay under the body of the spider, not quite dead, and was smeared with the filthy liquor or saliva exuded by the monster . . . the neighbors call them '*carangueijeiras*,' or crab-spiders. The hairs with which they are cloaked come off when touched, and cause a peculiar and maddening irritation. The first specimen I killed and prepared was handled incautiously, and I suffered terribly for three days afterward.

The Amazon *does* harbor an arachnid group capable of killing humans: the Brazilian wandering spiders (*Phoneutria* spp.), a species of which the *Guinness Book of World Records* once labeled "the deadliest spider on earth." In fact, the genus name—*Phoneutria*—is said to be derived from a Greek term meaning "murderess."

In Amazonia, wandering spiders do not spin webs to capture their prey but hunt them on the forest floor at night, seeking frogs, lizards, scorpions—even other spiders. During the day, they take refuge under rocks or fallen logs. When they feel threatened—for example, if a human turns over a piece of wood under which they have been hiding—they strike an immediate and unmistakable defensive posture with two front legs held high over their head, an arachnological version of

"Don't Tread on Me!"[19] Unlike many invertebrates who prefer to flee, these spiders are fast, fearless, and aggressive.

Wandering spider venom consists of a fiendish blend of peptides, proteins, and other toxins. A bite can prove incredibly painful and set off a cascade of additional unpleasant symptoms. While collecting plants in central Suriname, a colleague turned over a pottery shard under which one of these spiders was hiding. The arachnid bit him once on the ankle, and the botanist collapsed in pain. His heartbeat rapidly accelerated and became irregular (tachycardia); he suffered severe abdominal pain, nausea, blurred vision, hypothermia, and breathing problems. Fortunately, a nearby Amerindian shaman soon treated him effectively with plants and possibly saved his life. In other cases of wandering spider bites—particularly with children or the elderly—the result has been paralysis, asphyxiation, and sometimes death.

Bites by the wandering spider can also sometimes cause priapism: erections that can last for hours. One constituent of this venom—termed "Tx2-6"—is being investigated as a potential treatment for erectile dysfunction.[20]

What are Amazonia's most formidable predators?

The most fearsome Amazonian predator, one known to few outside tropical South America, is the Orinoco crocodile (*Crocodylus intermedius*). Reaching a length of more than 22 feet (7 m) and weighing over a ton (900 kg), this aquatic carnivore is a highly opportunistic feeder, eating everything within range, including birds, fish, snakes, turtles, caimans (including juveniles of its own species), and mammals (including, apparently, people). This crocodilian once ranged throughout the upper Orinoco Basin of Colombia and Venezuela, especially in the seasonally flooded llanos grasslands. It is also one of the world's most critically endangered reptiles: estimated at more than 1 million individuals just 80 years ago, the Orinoco crocodile population has been reduced to fewer than 1,500 individuals because of massive overhunting for the skin trade as well as because of habitat disruption—and the political chaos in Venezuela has resulted in even more crocodile hunting, both for meat and for leather.[21]

Nearly as large as the Orinoco crocodile, the black caiman (*Melanosuchus niger*) has a much greater range, being found throughout the Amazon Basin and into southwestern Guyana and eastern French Guiana. Black caimans devour anacondas (*Eunectes murinus*), armadillos, birds, capybaras (*Hydrochoerus hydrochaeris*), deer, fish (including piranhas), frogs, giant otters (*Pteronura brasiliensis*), lizards, monkeys, peccaries, sloths, snakes, tapirs (*Tapirus terrestris*), turtles, and other caimans. These giant crocodilians do not restrict themselves to solely native rainforest fare: they also consume cattle, dogs, horses, and pigs. And given that black caimans occupy a range more than five times larger than that of Orinoco crocodiles, they encounter humans much more often—and sometimes eat them. Henry Walter Bates published the classic account in 1863:[22]

> A large trading canoe arrived [at the town of Caicara] and the crew, as usual, spent the first day or two after their coming into port in drunkenness and debauchery ashore. One of the men, during the greatest heat of the day, when almost everyone was enjoying his afternoon's nap, took it into his head whilst in a tipsy state to go down alone to bathe. He was seen only by . . . a feeble old man who was lying in his hammock . . . who shouted to the besotted Indian to beware of the black caiman. Before he could repeat his warning, the man stumbled, and a pair of gaping jaws, appearing suddenly above the surface, seized him around the waist and drew him under the water. A cry of agony—"Ai Jesus!"—was the last sign made by the wretched victim. The village was aroused; the young men with praiseworthy readiness seized their harpoons and hurried down to the bank; but of course,

> it was too late, a winding track of blood on the surface of the water was all that could be seen. They embarked, however, in canoes, determined on vengeance: the monster was traced, and when, after a short lapse of time, he came to breathe—one leg of the man sticking out from his jaws—he was dispatched with bitter curses.

This was by no means an isolated incident: 23 fatal attacks by black caimans were recorded in Amazonia between 2000 and 2016. Considering that these creatures thrive in the most remote corners of tropical South America, it seems likely that additional killings went unreported.

Another sizeable reptile of concern to humans is the more famous anaconda. Though the anaconda is unquestionably the longest snake in the Amazon, most scholarly accounts conclude that the reticulated python of tropical Asia exceeds the anaconda in length, although the South American snake is much greater in bulk. Today, few anacondas exceed 20 feet (6 m), but Alfred Russel Wallace once wrote to fellow explorer Hamilton Rice in 1908 about a specimen whose "length of 38 feet [12 m] had been verified by careful and incontrovertible measurement." As humans move into ever more remote regions in increasing numbers, they typically destroy the largest specimens of predatory species, hence much larger specimens of anacondas, black caimans, and Orinoco crocodiles probably flourished in the past.

Like the boa constrictors (*Constrictor constrictor*) to which they are closely related, anacondas seize and constrict their prey.[23] Anacondas lurk below the water's surface near riverbanks with just their eyes and nostrils exposed as they wait for unsuspecting animals to approach and drink—the same technique utilized by both the Orinoco crocodile and the black caiman. Because of this practice, anacondas are seldom seen by humans, although an expert on the field behavior of the great serpents recorded two incidents in Venezuela in which anacondas pursued and attacked humans in a predatory rather than a defensive manner. He further argues that—despite many claims to the contrary—large anacondas are capable of swallowing humans. Nonetheless, there exist few accounts of anacondas attacking people, although the best is once again provided by the redoubtable Bates from Brazil:[24]

> The father and his [10-year-old] son went . . . a few miles up the Teffe to gather wild fruit, landing on a sloping sandy shore, where the boy was left to mind the canoe whilst the man entered the forest. . . . Whilst the boy was playing in the water under the shade of these trees, a huge [anaconda] stealthily wound its coils around him, unperceived until it was too late to escape. His cries brought the father quickly to the rescue, who rushed forward, and seizing the anaconda boldly by the head, tore the jaws asunder.

And yet another reason why there are so many fewer killings attributed to anacondas than to crocodilians is due to their method of attack. The victims of large constrictors are seldom able to cry for help as they are quickly wrapped in suffocating coils, and the remains of the victim are excreted weeks later. In contrast, victims of crocodilian attacks frequently cry for help, and the river is soon awash in blood.

Yet another dangerous creature is the largest carnivorous mammal of tropical America—the King of Beasts of the Amazon rainforest—the jaguar (*Panthera onca*). Ranging as far north as Arizona and as far south as Argentina, the jaguar is the world's third largest feline, exceeded in size only by the lion and the tiger. Males can tip the scale at more than 330 pounds (150 kg)—appreciably more than an African lioness. Jaguars are renowned for their exceptionally powerful bite: they can easily penetrate turtle carapaces (shells) and often kill mammalian prey with a single bite to the back of the neck.

Because of their power and strength, as well as their ability to roam the rainforest in the dead of night, jaguars serve as the symbol of the shaman—indeed, it is not uncommon for shamans to claim that, with the aid of powerful plants like ayahuasca, they can turn themselves into jaguars. Like the Orinoco crocodile, the black caiman, and the anaconda, these cats will attack and consume almost any vertebrate within reach. On rare occasions, they stalk humans.

In Manu National Park in Peru, a biologist once tranquilized a large jaguar to which she then attached a radio collar. One night, she was tracking this creature through the rainforest when she realized she was following the creature in a circle—the cat was in fact tracking her.

Jaguars tend to avoid most humans, especially when people are traveling in groups or—as is often the case with a hunting party—when they are accompanied by dogs. Prior to contact in the 1970s, the hunter-gatherer Akuriyo Indians of the northeast Amazon hunted alone and without dogs. Fully 30% of adult male Akuriyo hunters were killed by jaguars.

Are there hallucinogenic frogs in the Amazon rainforest?

In October of 1969, the intrepid photojournalist-explorer Loren McIntyre, mentioned earlier for finding one of the sources of the Amazon—was in search of a remote tribe in the northwest Amazon then known as the Mayorunas (now more commonly called the Matses) who were believed to live in the then uncharted forests near the Brazil–Peru border. They were also known as the Jaguar People because of their custom of inserting jungle grasses into their perforated nostrils to give them a catlike appearance in homage to the jaguar, their sacred animal. McIntyre managed to make contact with them near the headwaters of the Javari River and—at their invitation—followed them into the jungle.

After a nearly sleepless night in a Mayoruna encampment, McIntyre—again, at their invitation—joined in a dance that lasted several hours. At the conclusion of the festival, his tribal hosts opened palm leaf baskets and removed giant monkey frogs (*Phyllomedusa bicolor*). They then removed embers from the fire, applied them briefly to their forearms, then rubbed the amphibian skin secretions into the open wounds. Invited to participate, the American decided to partake and found the potion to be highly hallucinogenic.

According to McIntyre, the Matses employ the frog slime as "hunting magic": absorbing the amphibian secretions makes one smarter, stronger, bolder, able to communicate telepathically with other members of the tribe, and able to see precisely where prey could be found the next day. Subsequent laboratory analysis of green monkey frog skin yielded a host of novel peptides, at least seven of which are highly active in the human body. Two have been studied as a potential means of increasing the permeability of the blood–brain barrier, an important physiological challenge for delivering medicine to the brain even for modern medicine.

A perhaps even more striking find in the frog slime was a new opioid, termed *dermorphin*, which is more than 30 times more potent a painkiller than morphine. It seems also less likely to cause addiction, meaning that it might one day prove—in some form—to be useful both as a painkiller and/or as a treatment for addiction. Dermorphin has already demonstrated its utility in one peculiar, lucrative, and illegal way: doping racehorses. The compound has been administered to thoroughbreds to make them run faster without pain—and the horses can pass a drug screen designed to search for the presence of amphetamines, which dermorphin is not.

A better-known skin toxin is a dart poison which comes from Neotropical frogs. Though their habitat stretches from Central America down through tropical South America, most of the 170 species of the tiny creatures that harbor these toxins are found in the Amazon. The bright coloration—known as *aposematism*—is the herpetological equivalent of a warning sign to potential predators that these tiny and seemingly defenseless creatures are highly poisonous.

The most toxic of the poison dart frogs is used as an arrow poison by the indigenous inhabitants of the rainforests of the Choco in northwestern Colombia. There, they employ the skin secretions of *Phyllobates terribilis*, a 1-inch individual of which could produce enough poison to kill 10 men.

And another frog from this family, *Epipedobates tricolor*, from southwestern Ecuador, was found to contain a novel compound labeled "epibatidine" that is more than 100 times more potent than morphine but—in its natural form—is too toxic to be used in humans. As with dermorphin, a synthetic or semisynthetic form may one day prove useful in modern medicine.

Poison dart frogs are considered poisonous rather than venomous because they possess no delivery mechanisms with which to insert the toxin into the body of the victim. Researchers recently found that two species of frogs in northeastern Brazil possess unique—and uniquely deadly—toxins. These frogs—*Corythomantis greeningi* and *Aparasphenodon brunoi*—have bony spines protruding from their skulls that can pierce the skin of another creature—and their poison is more lethal than that of the dreaded bushmaster (*Lachesis muta*), the deadliest snake in the Amazon.

There are believed to be more than 1,000 species of frogs in Amazonia, and we are just now learning that many produce bizarre chemicals or extract them from the insects they eat and then sequester these compounds in their own skins to warn off potential predators. However, frogs are disappearing around the world due to their exceptional vulnerability to climate change, pollutants in the environment, and to chytrid fungi which disrupt their breathing and their nervous systems. How ironic that these "lowly" creatures are only now beginning to be appreciated as sources of new and potentially useful chemicals by the same species that bears most of the responsibility for destroying them.

Are there pink dolphins in the Amazon?

The Amazon harbors not one but three species of dolphins. The tucuxi or bufeo negro (*Sotalia fluviatilis*)—also known as the gray dolphin—is the smaller one, reaching less than 5 feet (1.5 m) in length. Gregarious and acrobatic, they usually travel in groups of two to nine and frequently leap out of the water. Tucuxis are common and widespread in Amazonia, often frequenting the mouths of rivers and streams and the bases of waterfalls. They range outside the Amazon as well, north into the Caribbean as far as Panama.

Much larger, much stranger, and more spectacular is the boto or pink river dolphin (*Inia geoffrensis*), which can reach almost 9 feet (3 m) in length and weigh 350 pounds (159 kg), three times heavier than the tucuxi. And what seemed to be a boto was recently determined to probably be a closely related but separate species (*I. araguaiaensis*) in the Araguaia River in southeastern Amazonia.

Unlike the gray dolphin, the boto avoids the ocean and thrives only in rivers of the Amazon and Orinoco drainage. The boto is also more of a solitary creature, often found alone or usually in groups of four or fewer.

Botos rank among the world's most extraordinary mammals in both appearance and capacity. Most are pink, with a long thin snout, tiny eyes, and a bulbous forehead (the so-called melon, which contains their biosonar apparatus for echolocation), strikingly reminiscent of a bald, sunburned, and nearsighted old man. Prodigious fish-eaters, they are known to consume more than 50 different species, including preferred game species like tambaqui (*Colossoma*) and pirapitinga (*Picractus*) as well as piranha and many catfish. Botos teeth are strong enough for them to catch, crush, and consume crabs and turtles as well.[25]

Having unfused neck vertebrae, they are extremely agile and are able to turn their heads at least 90 degrees, a distinct advantage when swimming into the flooded forest during the rainy season, where they often capture fish darting between submerged roots and tree trunks. Another peculiar habit they exhibit is swimming upside down, believed to be their means of searching for prey along river bottoms. And stranger still is their supposed propensity for sleeping upside down as well.

Many legends are associated with this creature. One persistent tale, common in Brazilian Amazonia, is that male botos can assume human form and come ashore for nocturnal forays, during which they seduce and impregnate local women. Ladies who find themselves dancing with a handsome stranger at Amazonian soirees are advised to surreptitiously stroke the scalp of their partner to check for a blowhole. More than a few unwanted pregnancies in Amazonia have been blamed on the allegedly perfidious boto.

Botos have enormous brains, possess no innate fear of humans, and frequently interact with people in a curious or playful manner. They sometimes retrieve thrown objects, grasp fishermen's paddles, and even grab a swimmer's hand under their flipper and tow them.

Near the town of Mocajuba in the Brazilian Amazon, pink dolphins having been playing with the children of the village in the river every morning for the past 30 years. According to Michael Goulding, cooperation between dolphins and fishermen in Amazonia was once common. For example, fishermen set weirs (traps) in the mouth bay of the lower Tocantins River. Botos chased fish into the weirs, and the fishermen rewarded them with some of the fish they caught.

Unfortunately, human–dolphin interaction in the Amazon is not always benign. Some fishermen in Amazonia kill botos to keep them from consuming fish, while others cut them up for fish bait, despite legal prohibitions against this in every country where they are found.

Botos may also succumb less directly to human behavior. When entangled in fishing nets, they may drown, being air-breathing mammals. Gold mining along rivers leaves poisonous mercury that is incorporated into the flesh of bottom-dwelling catfish, which form a significant part of the botos' diet. And dam building in the Amazon threatens the well-being of these wide-ranging river dolphins.

In recognition of their uniqueness, in 2012, Bolivian president Evo Morales declared botos to be "national treasures which could not be harmed."

Why does Amazonia harbor the most diverse freshwater fish fauna in the world?

The rivers of Amazonia contain the ichthyological equivalent of the Mos Eisley cantina in the Star Wars universe: a collection of bizarre and improbable creatures that sometimes defy belief. There are fish that spend most of their lives standing on their head (*Abramites*) and others that stand mostly on their tail (*Nannostomus*). There are fish that eat people (*Brachyplatystoma filamentosum*), fish that impale people (*Potamotrygon*), fish that shock people (*Electrophorus*), and fish that can invade people (*Vandellia*).

The Amazon harbors some of the world's largest freshwater fish (the 13-foot [4 m] pirarucu, *Arapaima gigas*, also known as the arapaima and the paiche); some of the smallest (less than an inch [2.5 cm] in length); some of the most beautiful (*Paracheirodon*); and some of the ugliest (*Plecostomus*). This ecosystem is home to hundreds of species of catfish, most of which have whiskers, as well as eight species of dogfish (Cynodontidae), all of which possess large canines.

The Amazon is home to fish that talk (*Acanthodoras*), fish that walk, fish with bony tongues (Osteoglossidae), fish with giant human-like molars (*Colossoma*), fish shaped like banjos (Aspredinidae), and fish that swim backward (*Gymnotus*). There are not only carnivorous piranhas—one species hunts in packs that can strip an ox in minutes (*Pygocentrus*)—but vegetarian ones (*Serrasalmus*), too, as well as fish that eat fish, fish that eat fruits, fish that eat seeds, fish that eat fins, fish that eat snails, fish that eat tails, and fish that eat scales.

The Amazon harbors fish that breathe air (swamp eels, *Synbranchus*), fish that live mostly in burrows in muddy soil (killifish, Rivulidae), and fish that fly (hatchet fish, *Carnegiella*). There are fish that climb trees and fish that hunt in trees: the arowana (*Osteoglossum*), also called the "water monkey" (*macaco d'agua*) because it can leap 6 feet (2 meters) out of the river to snag insects, amphibians, reptiles, birds, bats, and—reportedly—baby sloths and small monkeys from overhanging branches.[26]

The leaf fishes (*Monocirrhus*) convincingly mimic dead leaves floating in the water. When threatened, they sink to the river bottom in precisely the same manner as dead leaves. Most of their time is spent floating around aimlessly until an unsuspecting fish swims by, which is then quickly attacked and devoured.

One group of Amazonian fishes has eyes that are oriented downward to scan river bottoms, while the so-called four-eyed fish (*Anableps*) has two-chambered eyes that permit simultaneous aquatic and aerial vision as it skims along the surface. Other fish—usually living under cataracts or trees along riverbanks—have sharply upturned mouths. There exist fish that consume mostly mud, feces, flowers, freshwater sponges, wood (*Panaque*), mucus from their parents' skin (discus, *Symphysodon*), or chunks of other fish (*Serrasalmus*). Some catfish (such as the bacu pedra, *Lithodoras dorsalis*) are so heavily armored that they appear to be aquatic escapees from King Arthur's Round Table.

Noteworthy for the odd appearances and habits of its ichthyofauna, Amazonia encompasses stunning levels of fish diversity as well. As of 2016, the Amazon is known to harbor more than 3,000 species of fish, and dozens of new species are being found and described each year.

The Nile contains approximately 800 species of fish, the Congo has about 700, and the well-studied Mississippi is home to less than 400, meaning that Amazonia has more species than these three rivers combined. The Freshwater Fish Specialist Group of the International Union for the Conservation of

Nature (IUCN) estimates the total number of freshwater species to be approximately 15,000, meaning that one of every five freshwater species on earth inhabits Amazonia.

One peculiar aspect of this stunning diversity is that more than 80% of the species belong to a mere three groups: the Characiformes (including pacus, piranhas, and tetras, 37% of the total), the Siluriformes (catfish, also 37%), and the Perciformes (cichlids like the angelfish, the discus, the oscar, and the peacock bass, at 10%). Through adaptive radiation, the descendants of these ancestral groups of fishes—like Darwin's finches in the Galápagos—diversified and evolved into different species to fill different ecological niches. The fossil record demonstrates that the lineages that gave rise to modern Amazonian species are many millions of years old.

In addition, the presence of fossils closely related to current Amazonian species in adjacent areas like northern Venezuela or the Magdalena River Valley in central Colombia shows that landscape changes over time undoubtedly facilitated both speciation and extinction as populations were either joined or separated. Per ichthyologist John Lundberg (2001),[27]

> [t]he geological upheavals that divided rivers and river basins provided opportunities for speciation when fish populations were isolated . . . fossils from the regions now occupied by the Magdalena Basin and Lake Maracaibo [in Venezuela] show they once contained fishes no longer found there but that still inhabit the Amazon. . . . The biodiversity equation is balanced by speciation on one side and extinction on the other. Low extinction rates are part of the formula for fish species richness in the Amazon basin.

Many reasons have been put forward to explain the mega-diverse levels of Amazonian fish species. As mentioned earlier, the enormous size of Amazonia—more than twice that of the

Congo Basin—means that there are prodigious numbers of niches available for creatures to fill. Beyond the great forest, Amazonia also contains extensive, seasonally flooded grasslands. Additionally, strikingly different ecosystems are connected to the great rainforest along its western border, from species-rich cloud forests to Andean highlands, in addition to estuarine and oceanic ecosystems to the east. Then there is the Amazon Basin's open nature: it is both permanently connected to the Orinoco Basin to the north through the Casiquiare Canal and seasonally connected to the Guianas through inundated swamps in both western Guyana and eastern French Guiana.

Another factor is the latitudinal gradient phenomenon: in the wet tropics at low altitude, species numbers tend to increase as one approaches the equator. With more energy from the sun, equatorial ecosystems tend to have more food, heat, and sunlight and therefore more species. Since the Gondwanaland breakup, warm, wet, stable conditions have mostly prevailed over millions of years in northern South America. Other places, such as Alaska, Canada, and Greenland formerly enjoyed tropical climates and biota: palm trees, crocodilians, and turtles once thrived there, but northern and southern polar regions cooled over time. The Amazon suffers no winter die-off as happens seasonally in temperate zone. And Amazonia was never covered in species-killing glaciers during the Ice Ages.[28]

Is the tiny candirú catfish as terrifying as its reputation?

Of all the creatures encountered by the first European scientific explorers of the Amazon, none caused them more trepidation than the tiny candiru catfish.

The celebrated Amazon riverboat guide Moacir Fortes likes to begin his cruises with a lecture he labels "The Seven Perils of the Amazon River."

"The Amazon," he says, "is the world's largest, most magnificent, and most diverse river. It is home to thousands of species we enjoy, like the delicious pacu, the giant arapaima and the graceful pink dolphins we will see in the course of our voyage.

"But," he continues, dropping his voice for emphasis, "it is also home to seven species that enjoy *us*: the anaconda, the black caiman, the electric eel, the piranhas, the 12-foot [4 m], man-eating paraiba catfish, and the smallest and scariest of them all . . . [here he closes his eyes and shudders] . . . the candiru!"

These creatures are tiny: sometimes just 2 or 3 inches (5–7 cm) long. And candiru *are* bloodthirsty beasts: they slip behind the gill covers of larger fish—often long-whiskered catfish—and attach themselves to the gills, where they bite into and drink from blood vessels.

But this vampiric behavior generates only a negligible part of their reputation: in search of human blood, candirus are reputed to swim up the human urethra, where they then allegedly fasten themselves by spreading a series of backward-pointed spines, as if opening a tiny umbrella that then locks them in place, causing indescribable pain and sometimes death. The intrepid German explorer von Martius first reported this phenomenon in 1829, followed by an even more frightening account by the French biologist Castelnau who, in 1855, claimed that fisherman along the Araguaia River insisted that men urinating from the riverbank were running the risk of candirus swimming up the urine stream and into their bodies, thereby simultaneously defying the laws of both gravity and fluid dynamics.

Another early account described the candiru as "very small, but uniquely occupied in doing evil," whereas the BBC more recently reported this supposed peculiar predilection of the little catfish as "a horror story that is enough to keep your legs firmly crossed for days."

However, firsthand scientific and medical accounts of this candiru behavior are exceedingly rare. Small and parasitic fish in search of blood vessels would seem to have an easier time entering a human vagina or anus than ascending a urine column and then entering the end of a penis. And while there are somewhat credible reports of women being invaded by candirus who were attracted by menstrual blood, the purported cases involving men remain unverified.

Most accounts describe the candiru as belonging to one species—*Vandellia cirrhosa* of the Trichyomycteridae family—but it seems highly likely that several other species of this same genus exhibit similar behavior. *Vandellia cirrhosa* candirus have been featured and sensationalized in several recent television programs for the reasons just mentioned, but, for the most part, the media have neglected an equally interesting and much more verifiable story about other family members.

There exist a group of slightly larger candirus, belonging to the family Cetopsidae, sometimes referred to as the candiru-acu ("whale catfish") because of their odd, vaguely cetacean shape. They are piranha-like in their ferocity, occasionally going into a feeding frenzy that boils the surface of the water and that indeed is sometimes mistaken for a piranha attack. These candirus are known to bite chunks of flesh out of the legs of fishermen, and they often attack larger fish that have been hooked on a fishing line. As the unfortunate fish is being hauled in, the candirus may devour the underside, stomach, intestines, and back muscles of the living animal to the point where a fisherman may reel in little more than a head attached to a skeleton.[29]

How dangerous are electric eels?

On March 19, 1800, Alexander von Humboldt and Aimé Bonpland had departed Caracas, Venezuela, and were heading south toward the great rainforest. They stopped at the town of Calabozo, where they learned that electric eels thrived in nearby ponds and creeks. Ever the polymath, von Humboldt had vowed to study these creatures and the electricity they produced. Well before electricity was even remotely understood, the ancient Egyptians revered the electric catfish of the Nile.

After Benjamin Franklin's famous kite experiment in 1752, the scientific establishment hungered to learn more about electricity and the strange animals capable of producing it.

There are few creatures in the world as bizarre as the electric eel of the Amazon. These peculiar fish are essentially 8-foot (2.5 m) cylinders with flattened heads, some weighing more than 40 pounds (18 kg). Except when they are on the attack, they are sluggish creatures that swim and hover by means of an undulating fin along their belly that makes them look like a cross between a monorail and a helicopter. They obtain little of their oxygen from the water, and they surface regularly to gulp air.

The Trio Indians of Suriname regard electric eels as a favorite food and pride themselves on being able to lure these animals to the surface by means of a peculiar sucking-in whistle. Electric eels feed mainly on fish but are known to consume shellfish, amphibians, birds, small mammals, and fruits. Some of their favorite fruits are said to be those of the acaí palm (*Euterpe oleracea*), and there is an extremely widespread and persistent story in the Amazon that the eels are able to dislodge the fruit by "shocking" the roots of the palm tree. One often finds electric eels floating in swampy areas near these fruiting palms in a seemingly expectant manner.

These eels are effectively the most powerful natural Taser on the planet, capable of emitting bursts of 860 volts, five times the voltage of a standard US wall socket. Of course, not every electric discharge consists of maximum voltage. Electric eels have poor eyesight and employ low-voltage pulses to detect objects in cloudy water, much as bats use echolocation to find their way in the dark. The eels use stronger signals to force potential prey to twitch and reveal their hiding place. If the prey is deemed suitable, the eels will unleash a stronger current, thereby temporarily paralyzing the creature prior to consuming it. Given their ability to find and paralyze their victims from afar by invisible means, electric eels have been called "the Jedis of the Amazon jungle."

When von Humboldt informed his guides that he wished to collect and study the dangerous eels, the Indians replied that they would "*embarbascarlos con caballos*" ("stupefy them with horses"). Knowing that once the eels had discharged most of their electricity they could be handled with relative impunity, the Venezuelans rounded up 30 horses and mules and drove them into the swampy pool full of eels. Pure bedlam resulted as the eels rose out of the water as they blasted the unfortunate mammals with electricity: von Humboldt recounted how one eel "[pressed] all its length along the belly of the horse, giving it electric shocks." Two of the horses drowned, and the others were left weak and panting on the shore.

For more than two centuries, many researchers considered von Humboldt's account to be exaggerated as eels were not believed to rise out of the water to deliver their shocks. However, research by Dr. Kenneth Catania of Vanderbilt University in 2015 revealed that electric eels can and do propel themselves upward to administer high-voltage shocks to defend themselves against perceived predators. Von Humboldt's encounter with the eels in Venezuela took place in the dry season, in a pool from which the eels could not escape, and hence represented a defensive behavior that researchers may never have observed in a river or a stream. And by delivering the blow to the aggressor out of the water, the eels delivered the full impact of the electrical shock, which does not dissipate as it would if it had to first travel through water.

Nor was von Humboldt the first scientist fascinated by the electric eels of Amazonia: Carl Linnaeus, the father of scientific classification, first studied them more than 250 years ago. And Alessandro Volta was said to have designed the first battery inspired in part by the physiology of these odd creatures. Researchers are currently studying the eels in even greater detail in the hopes of improving the design of hydrogel batteries for powering medical implants.

Can electric eels kill people? The Wayana Indians of Brazil claim that these creatures generally prefer quiet water but also frequent the eddies behind rocks in river rapids because other fish congregate there to avoid being swept downstream by the current. Unfortunately, so do humans. While it is uncertain whether the eels can electrocute people to death, unlucky victims can be shocked into unconsciousness and may then drown.

Do piranhas attack humans?

Piranhas are widely reputed to be ravenous, ferocious, vicious, and terrifying predators that can rip to shreds and devour any creature that comes within reach. That these fish are so widely feared can be attributed to two factors: Teddy Roosevelt's bestselling 1914 book *Through the Brazilian Wilderness*, in which he described piranhas as "the most ferocious fish in the world" and "the embodiment of evil ferocity."

The second factor is that—on rare occasions, in certain circumstances—piranhas live up to their reputation.

As noted earlier, in 1914, Roosevelt and Brazilian explorer Cândido Rondon joined forces to travel the previously uncharted River of Doubt in the southern Brazilian Amazon. During the 7-week journey, 3 of the original 19 expedition members perished, and Roosevelt himself nearly died. At the time of the expedition, Roosevelt was ex-president of the United States and the most famous American in the world; thus his account became a bestseller. He detailed an event in which local peasants threw an old cow into a pond teeming with piranhas that had been starved for several days, and he reported that the swarming fish quickly reduced the poor creature to little more than a skeleton.

Though piranhas can be found as far south as northern Argentina, most of the more than two dozen species inhabit Amazonia. Not only do these fish have exceedingly sharp teeth, but recent estimates indicate that at least two piranha species have the most powerful bites of any living carnivorous vertebrate. Despite their dentition and reputation, many (if not most) tend to be omnivorous, and at least one species—the piranha mucura, *Serrasalmus elongatus*—primarily preys on the fins and scales of other fish. Piranhas consume mammals, birds, reptiles, fish (including their own species), insects, crustaceans, fruits, seeds, and leaves.

Amazonian peoples have long employed piranha jaws and teeth for a variety of purposes. The Kamayurá of the Xingu in the southeastern Amazon rub crushed medicinal plants on their arms and legs, and then scrape the herbal mixture into their bloodstream by scratching themselves with piranha jaws, a very painful practice. The Trio Indians of the Brazil–Suriname border region formerly scored arrowheads with piranha teeth so the curare-tipped heads would be more likely to break off in the prey's body.

Both indigenous and other local peoples are often blasé about the presence of piranhas, and they swim and bathe in these rivers with relative impunity. During much of the year, the most common injuries inflicted by piranhas on humans occur when a careless fisherman gets bitten by a fish being removed from a hook, when one is released into a boat and grabs a toe, or when a curious piranha takes a nibble of the foot of a swimmer.

The exception is during the dry season when piranhas (particularly the red-bellied piranha, *Pygocentrus nattereri*) become concentrated, hungry, and hyperaggressive in shrinking lakes. Local peoples give these places a wide berth—it is then that the piranhas' fearsome reputation is most assuredly earned. Michael Goulding has provided this account:[30]

> Unlike most of its other kin in the Amazon, the red-bellied piranha is often a group hunter, at least during the low water period when populations of both predator and prey become dense in those floodplain lakes. . . . Group size varies from about a dozen to a hundred individuals. Several groups, however, may converge in a feeding

frenzy if a large prey animal is attacked, though this is rare. Piranha groups move about in search of prey concentrations and, once located, they appear to spread out like a battalion of soldiers making ready for battle. Prey is usually ambushed by one or more of the group that act as scouts to pick out disadvantaged individuals. Once the prey is crippled, one of the individuals will latch onto it, and this is the main signal to encourage the others to join in on the kill. They attack at lightning speed, each taking a bite or two and then darting out of the feeding frenzy. If an individual becomes too gluttonous and continues to feed on the prey for too long, others will bite at its fins to chase it off so that they can have a chance at the kill as well.

Are there sharks in the Amazon?

In the initial 2 weeks of July 1916, four people were killed and one was injured by a large shark or sharks along the New Jersey shoreline. Some attributed the attacks to the great white shark (*Carcharodon carcharias*), but this is conjecture because it remains unclear whether the actual culprit was ever captured or killed. The final two killings were in Matawan Creek, upstream from the ocean. Matawan is mostly fresh and brackish water, a combination shunned by the great whites but frequented by another known people killer: the bull shark (*Carcharinus leucas*).

Bull sharks occur worldwide, can flourish in both saltwater and freshwater, and are found not just in the ocean but also in estuaries and even rivers. Fast and agile swimmers, they are often aggressive and unpredictable, and some posit that they account for most shark attacks on humans. Bull sharks can exceed 11 feet (3 m) in length and—pound for pound—have the most powerful bite of any cartilaginous fish known.[31]

Scientists long believed that Lake Nicaragua in Central America harbored an endemic species of bull shark. Subsequent research revealed that these "lake sharks" had each made their way upriver, salmon-like, from the Caribbean along and over eight sets of rapids in the San Juan River and into the lake, a distance of almost 120 miles (193 km).

Bull sharks are *diadromous*, meaning that they can survive in both freshwater and saline environments. Their kidneys, livers, and gills have evolved to maintain suitable salt and water ratios, a feat of which most marine fishes are incapable.

In the Amazon, bull sharks prey on fish, turtles, birds, dolphins, stingrays, and (supposedly) other bull sharks. They thrive in the fish-rich waters of the Amazon estuary. Some have been caught at Manaus, the great river port more than 800 miles (1,290 km) upriver from the town of Belém at the mouth of the Amazon. But a few specimens have ventured to the foot of the Andes, a journey of well over 3,000 miles (4,800 km). If a bull shark entered a hypothetical country-spanning river at Washington, DC, and headed due west, it would swim more than 300 miles (482 km) past Los Angeles.

Fortunately for swimmers and bathers, older and larger (and hence, more dangerous) bull sharks tend to prefer marine environments. Most bull sharks that ascend the Amazon measure less than 8 feet (2.5 meters) and seem uncharacteristically nonaggressive.

5

INDIGENOUS PEOPLES

When did the first humans arrive in Amazonia?

Deciphering the prehistory of ancient Amazonia presents a multitude of challenges. Bodies do not readily mummify in the ubiquitous heat, humidity, and constant precipitation. There exist no written records of pre-Columbian Amazonia that can be deciphered Champollion-like with some rainforest Rosetta Stone. In fact, there exists little workable stone in Amazonia. Even if there were, why would Indians bother with materials hard to lift and difficult to carve when they have tens of thousands of species of flowering plants that they masterfully cut, hollow, burn, weave, bend, twist, arrange, and tie into everything they need, from *malocas* (longhouses) to musical instruments?[1]

Who were the first inhabitants of Amazonia and when they arrived are two interlinked questions that are nestled into two bigger queries: Who were the first Americans and when did they appear? There exist many more clues for solving the latter questions, but they have generated much more controversy.

Beginning in 1926, excavations near Clovis, New Mexico, uncovered manmade spear points manufactured in pre-Columbian times. This led to the widely accepted "Clovis First" hypothesis that the manufacturers of these spear and/or arrowheads were the first inhabitants of the Americas who had walked across the Bering Land Bridge from Asia more than 13,000 years ago. The archaeological world was astonished in 1977 when Thomas Dillehay, then at the University of Kentucky, began excavating the Monte Verde site in southern Chile, which was more than 1,000 years older than the New Mexico finds.

Initially uncovered by local people widening an oxcart trail, Dillehay and his colleagues discovered an astonishing treasure trove of well-preserved remnants: carved wood (including house planks), berries, fruits, seeds, salt, artifacts of stone and ivory, mastodon meat and bones, and medicinal plants—including wads of boldo (*Peumus boldus*) mixed with seaweed, a mixture still prepared today as a medicinal tea by the local Mapuche Indians.

Up to that point, the conventional wisdom had been that the Americas had been settled by big game hunters from Asia carrying a distinctive type of stone spearhead called "Clovis points." They were believed to have trekked across the Bering Land Bridge and marched down to Tierra del Fuego at the far southern reaches of South America. That people were thriving in southern Chile more than 9,000 miles south of Alaska and making weapons without Clovis spear points 1,000 years before the Clovis culture led to a major rethink of how and when the Americas were settled. Because we now know that glaciers then blocked much of the central route southward through Alaska, the most accessible route was probably along the Pacific coast. The sea level was much lower, and it was a zone rich in fish, berries, marine mammals, and edible seaweed.

Hence, the first humans to reach the Amazon rainforest may have arrived from the west after having crossed passes in the Andes, though initial arrivals from the other three directions are not inconceivable, merely a bit less likely.[2,3]

Though dates of human occupation in excess of 50,000 years have been claimed for Serra da Capivara in northeastern Brazil and 20,000 for Chiribiquete in the Colombian Amazon, the two most widely accepted ancient sites in Amazonia are Monte Alegre in Brazil and Pena Roja in Colombia.

Found about 4 miles (6.4 km) north of the town of Monte Alegre on the northern bank of the Brazilian Amazon, the "Pedra Pintada" ("Painted Rock") site consists of caves adorned with striking paintings of humanoid stick figures, geometric designs, and handprints in brown, red, and yellow. Dr. Anna Roosevelt (great-granddaughter of former US president and Amazon explorer Theodore Roosevelt) and her Brazilian colleagues carried out much of the excavation in the early 1990s, unearthing stone spear points; the remains of fish, shellfish, birds, and reptiles; and pottery shards from some of the most ancient ceramics known from the New World. The inhabitants of the site were also consuming Brazil nuts (*Bertholletia excelsa*) and palm fruits, much as many Amazonian inhabitants still do today. Pedra Pintada is believed to have been first inhabited more than 10,000 years ago. Based on the wide array of plants and animals that comprised their diets, Roosevelt concluded that these early Amazonians were sophisticated and well adapted—meaning that these people were unlikely to have been the very first people who settled in this rainforest.

At the western end of Amazonia, Pena Roja is a 9,000-year-old site along the banks of the Caquetá, one of the largest rivers in Colombia. According to Santiago Mora, the Colombian anthropologist who led the excavations, the peoples who inhabited this site were consuming Brazil nuts and the fruits of more than 10 different palms, most of which, such as maripa (*Attalea maripa*) and buriti (*Mauritia flexuosa*), remain dietary staples of Amazonian peoples today. The original Pena Roja inhabitants made simple hoes to harvest edible, underground rhizomes and prepared stone tools to carve wood and to skin animals.

Mora wrote, "the inhabitants of Pena Roja are some of the earliest settlers of the northwest Amazon, however, they were not the first humans to enter the forest . . . the behavior of these inhabitants seems to suggest that they had an extensive knowledge of the region."[4]

Of course, the earliest sites in the Amazon (or elsewhere, for that matter) will always be difficult if not impossible to find, meaning that the first immigrants presumably arrived prior to the inhabitants of Pedra Pintada and Pena Roja. Anna Roosevelt has recently hypothesized that the earliest inhabitants of Amazonia were foragers who spread throughout the rainforest 13,000 years ago, a date that dovetails well with what is known about both Pedra Pintada and Pena Roja.

Meanwhile, DNA tests link most modern Native Americans to an Asian origin. However, we are learning that humans were much more mobile in pre-Columbian times than historians initially believed. Ethnobotanist Paul Cox and his students have proved that there was contact between South American and Polynesian peoples based on studies of the South American yam during pre-Columbian times. Based on archaeological evidence in northeastern Canada, we know that there was contact between Europeans and Native Americans prior to Columbus. And genetic analyses have shown puzzling commonalities between two Amazonian tribes—the Karitiana and the Surui—and indigenous groups in Australasia. In fact, many of the Surui seem to exhibit Melanesian facial characteristics.

As the last continent to be settled by *Homo sapiens*, South America will remain a high priority for scientists intending to better understand the details of human prehistory as our species spread across the planet. The best current hypothesis is that the first immigrants arrived in the Americas prior to 20,000 to 15,000 years ago, and the initial settlers appeared in the Amazon prior to 10,000 years ago. Given the ongoing revolutions in remote sensing and genetics, as well as an interdisciplinary research approach that incorporates and admixes fields as diverse as archaeology, human physiology, and linguistic analysis, a more detailed understanding of this prehistoric odyssey is certainly forthcoming.

How many people were living in Amazonia in 1492?

Accurately estimating population sizes of Amazonian cultures prior to the arrival of the Europeans presents many challenges. Unlike in Central America, the pre-Columbian peoples of South America left no written record. Rainforest bacteria, fungi, and insects devour organic materials, and the lack of suitable stone for construction means the archaeologist has a lack of durable material to scrutinize compared to colleagues studying mummies and pyramids in the deserts of Egypt or Peru. Moreover, the first European arrivals came in search of gold, slaves, and souls (in order of decreasing priority), meaning that the few chronicles that were recorded were not sensitive and insightful inquiries into indigenous cultures.

Despite the relatively meager data, researchers have attempted to piece together an estimate of population size and settlement patterns in pre-European Amazonia. One obvious starting point is the early chronicles, the earliest of which was by Gaspar de Carvajal, a Dominican friar who accompanied Francisco de Orellana down the Amazon River in 1541. Carvajal reported the presence of massive villages built on bluffs overlooking (and paralleling) the river, some of which extended for more than 18 miles and which he estimated to harbor a population of some 10,000 people. Some researchers have dismissed these estimates as wild exaggerations while conceding that these riverine locales offered a richer and more diverse resource base than did most of the rest of the Amazon, 98% of which is terra firme forest, markedly less productive in terms of food resources and therefore incapable of supporting enormous settlements like those recorded by Carvajal.

The proximate aquatic bounty enticed people to settle on or near these riverbank locales. Not only did the waters teem with more varieties of fish than that of any other river on the planet, but they also harbored a bonanza of other edible species that included thousand-pound (454 kg) black caimans, 800-pound (1,764 kg) manatees, and 200-pound (91 kg) side-neck turtles

(*Podocnemis* spp.) whose massive populations and oviposition could fill a beach with more than a million eggs over the course of just a few nights.

Living atop the bluffs not only offered protection from other potentially hostile tribes, it also kept these populations and some of their gardens from being inundated during the Amazon's annual flooding cycle, when it carries Andean sediments and deposits those in the surrounding *várzea*, the seasonal floodplain forests along whitewater rivers. And, over the course of thousands of years of human habitation, the composting of organic materials resulted in enormous stretches of *terra preta do Indio*—the famous "black earth of the Indians"—which enhanced soil fertility, thereby creating the most productive soils in the lowland Amazon.

However, most of Amazonia consists of poor soils. Because of this and the fact that most modern Amazonian tribes are relatively few in number, some believe that the original population of Amazonia was not much more than a million individuals. The lushest forest in the world emanating from a sandy and depauperate substrate was—in the memorable phrase of the late anthropologist Betty Meggers—a "counterfeit paradise."

Other research perspectives exist. Archaeologist Anna Roosevelt contends that these pre-Columbian Amazonian cultures were large, diverse, and stratified societies with sophisticated ceramics. Elsewhere, ethnobotanists like William Balee and the late Darrell Posey introduced the concept of the "anthropogenic forest." Their research in Brazil demonstrated that indigenous peoples in locales like Maranhão in the northeast Amazon and the Xingu in the southeast had long been managing the forest to increase the abundance of plants employed as foods and fibers. Recent estimates have hypothesized that somewhere between 12% and 40% of Amazonia may be anthropogenic. Such management could augment the utility of terra firme forests and support large populations by the planting of edible species like Brazil nuts (*Bertholletia excelsa*) and the expansion of prime hunting habitat (by converting forest to grasslands through controlled burning).

Geographer William Denevan spent six decades attempting to determine the pre-Columbian Amazon population level based on available evidence, using sources as disparate as conquistador chronicles and recent archaeological finds in Acre state. He has concluded that a total population of 5 to 6 million people is a reasonable approximation, although he notes that the number could theoretically be much higher if there exist extensive tracts of rich terra preta soils that remain undocumented but would have supported large settlements.

Perhaps the most convincing answer to the question as to whether the Amazon in 1492 was thinly inhabited pristine forest or a heavily populated cultural parkland was put forward in a 2011 article by Jos Barlow of Lancaster University and other Brazilian and British colleagues:

> [W]e suggest that the influence of historical peoples occurred along gradients, with high impacts in settlements and small and scattered Amazonian Dark Earths [terra preta], moderate impacts where enrichment planting occurred or where forests were affected by anthropogenic wildfires, and finally a large imperceptible footprint from subsistence hunting and resource extraction across vast tracts of Amazonian forests that are far from permanent settlements and navigable rivers.[5]

What are the indigenous languages in Amazonia, and why is it important to study them?

Amazonia's extraordinarily high levels of biodiversity are widely celebrated; less well known is that this rainforest harbors extraordinary linguistic diversity as well. Dr. Alexandra Aikhenvald of James Cook University estimates that lowland Amazonia harbors more than 350 languages, most of which can be grouped into 15 families, although there are many isolated tongues. The only region that harbors a greater number of languages is New Guinea, with its many isolated valleys giving birth to a multitude of idioms. Amazonia, by comparison, is relatively flat, hence the proliferation of so many languages throughout lowland South America is relatively puzzling. As is the case with biological conservation, scholars are in a race against time as local languages disappear.

Most indigenous languages in Amazonia can be classified as belonging to one of four major groups: Arawak, Carib, Macro-Je, and Tupi. There are many isolated languages that do not appear related to these language groups.

The Arawak group represents the most widespread and contains the highest number of languages: at least one Arawak language is spoken in eight of the nine Amazonian countries. Arawak languages are found from the Atlantic coast of the Guianas in northeastern South America west to the Colombian Amazon, and south of the Amazon, originally as far as Argentina.

The second most numerous of these major groups is the Carib family, from which we derive the words "Caribbean" and "cannibal." As with Arawak, there are Carib speakers on the Atlantic coast of the Guianas, and the language group once ranged as far west as Chiribiquete in the central Colombian Amazon, where lived the warlike (and Carib-speaking) Karijona who are today almost extinct. Some Carib-language speakers live in southeastern Amazonia along the Xingu River. Most Carib speakers live in French Guiana, Guyana, Suriname, and adjacent Brazil.

Macro-Je speakers are by and large centered in the southern Amazon. Isolates of the Tupi grouping are found throughout the Amazon. However, the highest concentration of Tupi languages is in southwestern Amazonia, primarily on the upper reaches of the Madeira River in the Brazilian state of Amazonas.

One of the many reasons for studying and documenting Amazonian languages is to help determine and better understand the prehistory of the region. As mentioned earlier, the ancient history of this rainforest region is particularly difficult to decipher for several reasons: little fossilizes or mummifies in the wet tropics; the ancestors of today's Amazonian inhabitants created tools, weapons, and buildings from plants rather than stone; and these peoples never created written forms of their languages. Better understanding how many languages there are and were, how they are related, and when they split should help us determine a better timeline and understanding of how and when Amazonia was populated.

Unfortunately, relatively little is known about or documented for the majority of Amazonian languages—thorough studies of most languages are few and far between. Moreover, many languages completely disappeared with the advent of the Europeans. The first European to sail the entire length of the Amazon was Orellana in 1541, but the first serious linguistic studies of local indigenous peoples were not carried out for another 200 years. In the meantime, introduced diseases wiped out entire tribes, and enslavement and outright murder by European invaders was widespread. Today, linguists lament philological "black holes" on Amazonian maps that provide the names of long-extinguished tribes but not their languages.[6]

At least two intriguing trading languages thrive in the Amazon today. *Língua Geral*, also known as *Nheengatu* ("good speech"), is a lingua franca originally based on the Tupinambá indigenous language, which once flourished primarily along the Brazilian coast. Missionaries promoted it as a trading language so that different tribes and Europeans could communicate. Língua Geral is still widely spoken in the Brazilian Rio Negro.

Sranan Tongo is a very different trading language found primarily in Suriname but also in the border regions of neighboring Brazil, French Guiana, and Guyana. It is an English-based creole using Dutch, English, and Portuguese with some words derived from both West African dialects and Yiddish (some Maroon slaves escaped from plantations owned by Jewish merchants, and they had learned some of the owners' European languages).[7]

Today, the most widespread languages in Amazonia are Portuguese and Spanish since these are the national languages of the six largest countries: Brazil, Bolivia, Colombia, Ecuador, Peru, and Venezuela. Somewhat surprisingly, many Amerindians speak English in Guyana (formerly British Guiana), Dutch in Suriname (formerly Dutch Guiana), and French in French Guiana. Popular culture and messaging are usually communicated in the national language and thereby contribute to a declining fluency in traditional idioms among younger members of the forest tribes.

What is a shaman?

A shaman fulfills many roles in a tribal society, but first and foremost, he—or more rarely, she—is a healer. The shaman may be expected to diagnose, treat, and cure disease; predict and control the weather; interpret dreams; bring good luck in the hunt; provide protection against evil spirits sent by rival shamans; enhance fertility; prevent famines; and be a singer of songs and keeper of legends. He therefore serves as physician, psychiatrist, pharmacologist, priest, astrologer, weatherman, and *psychopomp*—the person who conveys souls to the underworld. In Amazonia, many tribes have a unique term in their own language for a shaman, but the most common regional terms are *paje* or *payé*.

The basic shamanic precept is that there exists both a connection and a balance between the physical world and the invisible realm. The shaman mediates between the two and is the only member of the tribe or village who has the knowledge and the courage to do so.

In the shamanic system, disease may originate from a variety of causes: evil spells, breaking of a taboo, bad luck, lack of protection, or even invasion of the body by invisible creatures that rapidly replicate (by analogy, perhaps, a reference to bacteria or viruses). To a skilled shaman, most diseases have obvious manifestations and straightforward treatments, usually direct application of plant material to the affected area (e.g., for skin diseases) or ingestion of a plant-based potion (e.g., for digestive problems). In addition to sometimes astonishing botanical expertise, a single shaman may know and employ more than 100 different plants for medical purposes. These healers may also utilize a wide variety of techniques increasingly incorporated into Western medical practice: aromatherapy, dietary additions or restrictions, hypnosis, kinesiology, massage, and other modalities of stress management.

When shamans are presented with difficult cases—especially ailments that appear to have an emotional or spiritual component—they believe they must consult and/or visit the spirit world to learn the origin of the malady and to discern how it can best be treated and cured. This process can be direct, as by purposely dreaming about the patient, or it may necessitate specific steps, as when entering a waking trance state. Some shamans access this state by smoking enormous tobacco cigars, others by the rhythmic repetition of the shaman's drum or maraca (rattle) or dance, still others by powerful plant magic: hallucinogenic snuffs (yopo, *Virola* spp., in the northern Amazon) or hallucinogenic potions (ayahuasca, *Banisteriopsis caapi*, in the western Amazon).

Having accessed the other world, the Amazonian shaman is said to receive assistance from spirit helpers, whether a deceased ancestor and/or the shaman's mentor, a totemic animal (usually a jaguar or a harpy eagle), a plant goddess (usually the spirit of tobacco or ayahuasca), or a master spirit of local animals (*el dueno de los animales*). These spiritual colleagues are perceived to assist the shaman in a variety of ways: to reach a correct diagnosis, to learn a cure, or even to transport the shaman to where the cure can be found by carrying him through the rainforest (in the form of a jaguar) or over the canopy (in the form of an eagle).

Almost all South American plants and plant compounds that found some role in Western medicine—such as tolu balsam, curare, coca, pilocarpine, and quinine—at some point were originally employed by tribal societies. And in-depth studies of Amazonian shamans are revealing intriguing results: that some of these shamans have sophisticated diagnostic abilities; that some also employ frogs, fungi, and insects for healing purposes; and that they may possess and employ treatments for intractable diseases, like AIDS and cancer, that are yet to be fully evaluated in the laboratory. While at least some of their healing approaches remain difficult to test, quantify, and reproduce in the lab, new technologies are making it far easier to find, evaluate, and potentially employ the chemical bases of these treatments than ever before.

Are shrunken heads fact or fiction?

Many warrior cultures have kept or used the bodies (or body parts) of dead enemies for various purposes. The Romans crucified Spartacus and 6,000 members of his slave army as a warning to other would-be rebels. The ancient Scythians used the hollowed-out arms of their vanquished foes as quivers for their poison-tipped arrows. And numerous Eurasian tribes employed the skulls of their foes as wine chalices, celebrating their triumphs in a most macabre manner.

Headhunters of the Amazon are more common in the public imagination then they ever were in the South American rainforest. Only two very separate groups were known to take heads since the Conquest: the Munduruku of central Brazil and the Jívaro (now more commonly known as the Achuar, Aguaruna, Huambisa, and Shuar) of the westernmost Amazon, mostly in Ecuador. Unlike the Munduruku, they not only decapitated their enemies but also shrank the separated heads. The practice was supposedly halted in the 1970s.

The earliest accounts of headshrinking date back to the 1500s, but widespread Western fascination with the practice began with explorers' accounts reaching a broad audience in the 19th century. Perhaps the undying appeal of this custom is analogous to that of mummies, in that one achieves a peculiar form of immortality. Western cultures' fascination with shrunken heads is longstanding, with the motif appearing in popular narratives from *Moby-Dick* in the 19th century to *Beetlejuice* in the 20th to *Pirates of the Caribbean* in the 21st. According to biologist James Castner, an authority on the topic of shrunken heads, "Their creation fascinates some, repulses others, but intrigues almost everyone."[8]

These indigenous peoples were and are known for their fiercely independent and bellicose nature. Not only was the great Inca Empire unable to conquer them, but the Spanish failed to subdue them as well. Obsessed with gaining access to the gold in Jívaro territory, in the 19th century, Spaniards entered into trade relationships with the Indians. Insatiable Spanish greed eventually doomed the agreement, and these indigenous peoples rebelled, massacring more than 10,000 Spaniards in what is considered one of the most forceful and effective Amerindian retaliations against European abuse and exploitation ever.

Most of the shrunken heads (known in local dialect as *tsantsas*) they made, however, were not simply war trophies: they were harvested and prepared for reasons of belief, spirituality, and power. Raiding parties could number from several hundred to less than three people. Rather than a huge pitched battle, raids tended to be carried out as ambushes in the belief that taking and shrinking an enemy head was a means of harnessing warrior spirits while preventing the soul of the deceased from wreaking vengeance.

Once the head had been excised, the skull was removed through an incision at the back. Flesh and fat were scraped away, and the head was boiled to separate out the remaining grease and convert the trophy into a leathery mask. They then carried out a complicated series of procedures, including filling and emptying the head repeatedly with hot sand and heated rocks to shrink the specimen while retaining something of the victim's original features. Both the eye and the mouth openings were sewn shut to maintain a head-like shape and to trap the original spirit inside. During and after the preparation of the *tsantsas*, feasts were held to transfer the power from the *tsantsa* to the warrior, to protect the warrior from the avenging soul, and to celebrate the victory.

And their reputation for shrinking heads helped keep outsiders at bay: in the 1940s, Norwegian explorer Thor Heyerdahl had trouble finding guides in Ecuador to lead him into the rainforest in search of balsa trees (*Ochroma pyramidale*) so great was local fear that the indigenous tribes would decapitate them and shrink their heads.

Museums, curio collectors, and tourists created an accelerating demand for shrunken heads. When the tribespeople realized that a *tsantsa* could be exchanged for a shotgun or rifle, local killings are said to have increased sharply. At the same time, morgues as far north as Panama were alarmed and puzzled that heads were sometimes being removed from corpses for unexplained reasons. Meanwhile, other tribes and even peasant peoples began fabricating surprisingly convincing versions of *tsantsas* from the heads of both monkeys and sloths.

It has been estimated that as many as 80% of *tsantsas* in museums and private collections are fake. Recently, however, the Israelis have developed a process for extracting and identifying the DNA from shrunken heads. The very first *tsantsa* examined was revealed to be not the head of a monkey or a sloth—but that of an Ecuadorean man.

What is slash-and-burn agriculture?

As we have seen, Indians who lived along the *várzea*—the seasonal floodplain forests inundated by whitewater rivers that deposit nutrient-rich Andean silt—flourished on both the rich riverine resources (particularly fish and side-necked turtles of the genus *Podocnemis* spp.) and the abundant crop yields produced on the fertile soils. It was in these zones, mostly on or near the main stream of the Amazon, that the early explorers observed massive Amerindian villages that sometimes extended for several miles. The vast majority of Amazonia, however, is upland forest—the so-called terra firme—characterized by soils that are both mineral- and nutrient-poor. Eking out a living in such a depauperate environment proved considerably more challenging than in the infinitely more fecund *várzea*.

Deciding how best to release the nutrients bound up in the vegetation into the infertile soil was the challenge. One solution was to fell the forest, allow the vegetation to desiccate for months in the tropical sun, then set it ablaze—hence the term "slash-and-burn" agriculture. The fire kills fungi, insects, nematodes, parasites, pathogenic bacteria, and weeds. The burning reduces all but the largest tree trunks to ash and releases the minerals and nutrients—particularly calcium, magnesium, nitrogen, phosphorus, and potassium, all extremely beneficial for plant growth—thereby rendering the soil (temporarily) fertile. The ash also raises the pH of the extremely acidic soil, further benefiting the crops being planted.

In Amazonia, the productivity of these gardens tends to decline sharply after 4 or 5 years, and the plots are typically abandoned after about 7 years as the Indians move to another site and restart the process. Traditionally, the abandoned gardens lay fallow for 20 to 100 years or even longer prior to being felled again. During the fallow period, woody secondary growth—bushes, lianas, and fast-growing light-adapted pioneer tree species—thrive in the abandoned garden. Nutrients accumulate in this pioneer vegetation, soil porosity and invertebrates return, and botanical species diversity slowly begins to increase. Such a slash-and-burn system could continue indefinitely if human population density remains low, people are semi-nomadic, and fields are left fallow long enough to eventually be farmed once more.

Economic botanist Charles Clement estimates that Amazonian indigenous peoples developed uses for about

5,000 local plant species prior to 1492, and more than 80 of these were selected for propagation. A significant number have been domesticated and are grown not only in Amazonia but in other parts of the world as well, including achiote (lipstick tree, *Bixa orellana*), cacao (chocolate, *Theobroma cacao*), coca (*Erythroxylum*), guava (*Psidium*), manioc (*Manihot*), papaya (*Carica papaya*), peach palm (*Bactris gasipaes*), peanut (*Arachis hypogaea*), pineapple (*Ananas comosus*), sweet potato (*Ipomoea batatas*), tobacco (*Nicotiana tabacum*), and yam (*Dioscorea*). Other Amazonian crops planted in pre-Columbian times also seem destined to be cultivated overseas, including acaí (*Euterpe oleracea*), cupuaçu (*Theobroma grandiflora*), guarana (*Paullinia guarana*), and guama—also known as the ice cream bean (*Inga edulis*).[9]

In global terms, the most important crop of Amazonian origin is cassava. Domesticated in the southern Brazilian Amazon and adjacent Bolivia more than 8,000 years ago, it is a woody shrub that produces a starchy tuberous root. Cassava is the world's sixth major food crop and the main carbohydrate food source for more than 750 million people, not only in South America but also throughout the tropics. Based on the percentage of cyanogenic glycosides contained, it is classified into two primary varieties: sweet or bitter. Sweet cassava typically requires cooking to detoxify the harmful compounds; bitter requires extensive handling and preparation before it can be safely consumed.[10]

In traditional Amazonian societies, the men typically clear and burn the gardens, while women do the planting, harvesting, and cooking. A very typical scene and sound in an Amerindian *maloca* (roundhouse) is that of the women and their daughters peeling and grating cassava roots, which are then stuffed into a tube (*matapi* in the northeast Amazon; *tipitipi* in the northwest Amazon) woven from arrowroot (*Ischnosiphon arouma*) or jacitara palm (*Desmoncus polyacanthos*) fibers and hung from the rafters of the hut. When downward pressure is placed on the tube's contents by inserting a pole through the lower end, a liquid containing the cyanogenic glycoside is squeezed out. The dried remnant can be used to make a beer or sifted into a flour to make porridge or bread, while the liquid itself is boiled to detoxify it and create a savory sauce known as *tucupi*.

Indigenous Amazonians maintain an astonishing diversity of cassava varieties: more than 200 varieties are grown by the Amuesha people in the Peruvian Amazon, while anthropologist Janet Chernela recorded 135 different types in four gardens of the Tukano Indians in Colombia. The Jívaro of the Ecuadorean Amazon tend more than 100 varieties, while 50 different cassavas are known from the gardens of the Kuikuru of the Brazilian Xingu. Unlike the monocultures cultivated on massive corporate farms, crop diversity found in indigenous plots acts as a hedge against insect infestation and a changing climate.

Nonetheless, the viability of the slash-and-burn system remains at best questionable in today's Amazon. Population density is on the rise, meaning that increasingly less pristine forest is available. Whereas Brazilian caboclos and other Amazonian peasant cultures that are adapted to the rainforest practice a system similar to that of the Indians, peasants newly arrived from the Andes or fleeing poverty and the lack of economic opportunities in cities like Bogotá or Rio de Janeiro have little comprehension of how to best farm the forest. Not knowing how to select the best sites, relatively ignorant of the optimal timing for felling and burning of the vegetation, and employing the wrong varieties of manioc and other crop species (corn, pineapples, papayas, and yams), they also apply fallow periods that are too short in duration to ensure food security over the medium to long term. Add to this that rural peoples—both peasant and indigenous—show less inclination to relocate from a given rainforest locale after a few years, preferring proximity to schools and cell phone towers, and the outlook for the continuing feasibility and practice of traditional slash-and-burn agriculture across the Amazon seems poor.

Do uncontacted and isolated tribes still exist?

The concept of uncontacted or isolated rainforest tribes enthralls many people, even those who have no other particular interest in tribal cultures or rainforest conservation. Ten years after they were published, people still recall pictures of red- and blue-painted Indians along the Brazil–Peru border attempting to keep the outside world at bay by firing arrows at a plane overhead.

Are these Indians truly uncontacted, or are they merely isolated? It seems fair and reasonable to rule out the existence of "undiscovered" or "lost" tribes. Anyone fortunate enough to live and work with relatively unacculturated indigenous peoples in the Amazon well knows that they are geographic masters of this challenging landscape—they do not get lost. And the question of "undiscovered" needs to be reasked: Undiscovered by whom? Columbus may have gone to his grave believing that he "discovered" America, but the Indians who were there to greet him would not have agreed.

Crucial to the discussion, therefore, is defining the term "uncontacted tribe." If the last Karijona *maloca* (longhouse) in the Chiribiquete region of the northwest Amazon was seen in 1904, and longhouses are once again being spotted there from overflights in small planes, can we say with any degree of confidence that these people are Karijonas and that they therefore do not represent an uncontacted group? If a Colombian missionary met a tribe of isolated Indians near the Yari River more than 20 years ago and presented them with machetes and pots, and they then fled and he was unable to find them again, can we say that they have been contacted? If a single Trio Indian had a single encounter with two previously undocumented tribes, are those tribes considered "contacted"?

Many specialists prefer to refer to these people as "isolated" tribes, which avoids the semantic complications. And they live in isolation by choice: they are well aware that an outside world exists, but they choose not to join it.

One might wonder why these tribal peoples choose to remain in isolation. It may be an adaptive response to the many epidemics that swept through the New World with the arrival of the Europeans in 1492. In the western Amazon, however, this retreat into the deepest depths of the rainforest was to escape the depredations of the Rubber Boom, beginning around 1900.

The record of Westerners' initiating contact to "help" or "save" these tribes has (for the most part) been a sad and sorry one. Fundamentalist American missionaries initiated contact with the Akuriyos of Suriname in the 1960s: within 2 years, one-third of the tribe was dead, including most of the people over 40 years of age. In a preliterate culture, the elders are the repositories of all tribal wisdom, knowing which plants are edible and which plants heal. Similar stories abound of tribes brought "in from the wild" who then take to their hammocks and die of respiratory diseases, depression, and/or culture shock.

Although there are recent reports of isolated tribes or groups from every Amazonian country except French Guiana and Guyana, the epicenter of these isolated groups is the northwest Amazon, where Brazil, Colombia, and Peru meet. There, dozens of isolated groups live—or have lived, until recently—relatively unmolested by the outside world. In an increasingly globalized planet, however, destructive forces are literally pressing in from all sides: extractive industries, narcotraffickers, unscrupulous ecotourism operators, and even fundamentalist missionaries.

The way to best protect these cultures and honor their wish to be left alone is simple in theory and fiendishly difficult in execution. First and foremost: protect their rainforest homes and keep outsiders out. Empower neighboring tribes as forest rangers to keep the outside world at bay. Partner with nearby peasant communities to engage their assistance as well. Enact and enforce relevant legislation. And train and finance knowledgeable and experienced medical personnel to be ready when contact does occur, either by accident or purposefully, if the isolated groups emerge seeking assistance against the introduction of imported disease or asking for assistance to evict illegal squatters, miners, loggers, or narcotraffickers who have encroached on their lands.

6

HISTORY

THE STRUGGLE FOR THE AMAZON

What was the Treaty of Tordesillas?

Signed in a town 100 miles (161 km) northwest of Madrid on June 7, 1494, the Treaty of Tordesillas marked an important milestone in the history of the Amazon. At the time, Portugal and Spain were the world's two major seafaring powers, and the treaty was an attempt to divide the newly discovered lands of the Americas between them. An imaginary line was drawn 370 miles (595 km) west of the Cape Verde islands—roughly the longitude of present-day São Paulo. Everything east of the line "belonged" to Portugal and everything west to Spain.

The notorious Renaissance pope Alexander VI oversaw the proceedings.[1] A Spaniard by birth, he believed that the arrangement proposed in the treaty granted Spain the most lucrative lands: China and India. However, the treaty was signed only 2 years after Columbus's first voyage, at a time when accurate geographic knowledge was laughably thin. The European estimate of the size of the world at that time was based on Ptolemy's calculations made more than a thousand years earlier. Unaware of the existence of the Pacific Ocean, Columbus called all the local people he encountered "Indians," and he believed that the island of Cuba was what Marco Polo called "Cipangu" (Japan).

The aforementioned provisional dividing line allocated almost the entire Amazon Basin to the Spanish empire. The Spanish, however, spent their primary energies pillaging and destroying the Aztec Empire in Mexico and the Inca Empire in Peru, not in Amazonia.

Being maritime nations, both Portugal and Spain focused their colonial efforts in South America closer to the east and west coasts, respectively. Meanwhile, the journeys of Orellana and Aguirre—both Spaniards—down the length of the Amazon in the mid-1500s failed to uncover evidence of the gold that so obsessed the Europeans. The sheer ignorance of cartography married to the towering arrogance of claiming other peoples' lands (first and foremost, that of the indigenous peoples themselves who were never part of the discussion) led to continuous jockeying back and forth as the Portuguese and the Spanish tried to expand their stakes. Ultimately, the Portuguese proved more aggressive, moving west *up* the Amazon—primarily for slaving raids—than the Spaniards did moving east and *down* the Amazon. When the 1750 Treaty of Madrid finally settled the dividing line between the Portuguese and the Spanish, the canny Portuguese had taken control of 70% of Amazonia and took possession of a country—Brazil—nearly 100 times larger than Portugal.

Who were the conquistadors, and why was Lope de Aguirre considered the worst?

The conquistadors were—for the most part—savage, bloodthirsty, greedy, cruel, inhuman men. But among that monstrous crew, Lope de Aguirre stands out as perhaps the most wicked of all. John Hemming—the preeminent historian of the Amazon—assessed him as "a man of unmitigated evil, cruel, psychopathic and gripped by an obsessive grievance against the whole of Spanish society." In his classic *Explorers of the Amazon*, Anthony Smith wrote that "even in a period of butchery and conquest, of rebellion and perfidy, [Aguirre's] exploits were in a class of their own."

Many of the leading conquistadors—like Hernán Cortés, conqueror of the Aztecs, and Francisco Pizarro, conqueror of the Incas—hailed from the exceedingly impoverished Extremadura region of southwestern Spain. Aguirre, however, was born in the wealthier Basque province in northern Spain near the French border. And—unlike many of the men with whom he conquered tropical America—he had some noble blood, meaning that (in all probability) he looked down on his lowborn (but ultimately wealthier) colleagues. Furthermore, under the primogeniture system, all family inheritance would belong to the eldest brother, indicating that Aguirre would be left destitute by his family. These factors combined to contribute to an outsider personality in Aguirre that made him forever resentful, paranoid, and vengeful. In this Basque warrior, however, these traits and his personal history united to generate a monster whose bloodlust was seldom equaled in the history of the Conquest.

The unparalleled riches found and stolen by Cortes in Mexico and Pizarro in Peru inspired generations of Spaniards and other Europeans to sail for the New World in search of similar treasures. But by the time Aguirre arrived in Peru in the mid-1530s, there were no more fabulous kingdoms to be found and sacked. Consequently, many of the conquistadors turned on each other, engaging in seemingly interminable protracted battles and civil wars. Between fighting (essentially) pointless skirmishes for and against the Crown in Peru, Aguirre also sailed and hiked in search of booty to Nicaragua, Bolivia, and Ecuador. After breaking laws, being found guilty, pursuing the judge for 3 years, and then murdering him, Aguirre was scarred mentally and physically: first from a severe public flogging, then by losing a hand in a battle with the indigenous peoples, and finally by two musket shots that left him with a permanent limp, further fueling his anger, pain, and paranoia.

Meanwhile, the conquistadors had spread from Colombia south to Chile without encountering kingdoms anywhere as rich in gold and silver as those of the Aztecs and Incas. The only unexplored land lay to the east: the Amazon rainforest. The conquistadors assumed that this great wilderness, teeming with mystery and unknown tribes, must harbor empires of monumental wealth. In 1560, the Spanish nobleman Pedro de Ursua—representing the Spanish Crown and accompanied by 300 Spaniards and many more indigenous peoples from the Andes—was ordered into the Amazon in search of this phantom treasure. Some historians have speculated that the Spanish rulers had an additional motivation to encourage expeditions into the Amazon: idle conquistadors were a violent and rebellious lot, and only exploring and conquering new frontiers could exhaust this dark energy.

German director Werner Herzog's 1972 movie, *Aguirre: The Wrath of God*, unquestionably represents one of the most accurate cinematic depictions of the Conquest ever captured on film. The stunning opening—conquistadors in full armor, accompanied by horses, mistresses, and highland tribespeople, slowly descending through the Andean cloud forest into the Amazonian lowlands—brilliantly captures the suicidal stupidity and extreme folly of this doomed expedition.

In reality, as reflected in the film, the Andean Indian porters knew little more about Amazonia than did the Spaniards, and taking horses into rainforests where there were no trails was monumentally foolish. Moreover, the interlopers wildly underestimated the amount of food they needed to feed such an enormous contingent, and neither the Spaniards nor these indigenous peoples knew how to hunt and fish in the rainforest. Worse, they often traveled along whitewater rivers infested with biting insects, meaning that—in that age prior to the invention of insect repellent and mosquito nets—they must have been driven to the point of madness from insect bites. And the Spaniards' predilection for wearing body armor combined with their utter lack of personal hygiene undoubtedly resulted in agonizing skin ailments in the constant heat and humidity. As if this was not punishment enough, both the indigenous porters and the Spaniards quickly began to perish from hunger.

As Herzog portrays him, these stresses further unbalanced the already mentally unstable Aguirre, who then launched an internecine campaign of slaughter. As they made their way east through the forest and along the great river, Aguirre had Ursua—the personal representative of the Spanish government—murdered in a bizarre bid to launch a conquest of Peru. Dissembling his intent, he had another soldier proclaimed prince of Peru and then ordered him killed in turn. Anyone he suspected of insubordination was immediately slain as they made their way down the Amazon. All the while, the Spaniards attacked villages, killing the inhabitants and stealing their food. Finally, Aguirre seized full control, forcing his men to pledge loyalty to him on point of death. Meanwhile, Aguirre had murdered more than a quarter of his fellow Spaniards.

The chronicle of this expedition is frighteningly detailed in parts and frustratingly vague in others. Historians debate whether Aguirre and his increasingly small band of cutthroats reached Venezuela by traveling north up the Rio Negro, through the Casiquiare Canal and down the Orinoco, or down the Amazon and then northwest along the South American coast. Based on Orellana's account from several decades earlier, the Spaniards knew they could reach Venezuela and the Andes via the Amazon route; there was no similar well-known account for the Rio Negro route.[2]

The Spaniards reached the Atlantic and sailed northwest to Margarita Island, where they launched an orgy of raping, looting, and killing. From there, they set sail south for Venezuela, where they were soon surrounded by Spanish troops loyal to the king of Spain. Before perishing, Aguirre stabbed his daughter to death so that no one else could have her. Many of Aguirre's compatriots turned on him, and he was killed, drawn, and quartered. His head and pieces of his corpse were publicly displayed as a warning to others.

Herzog's film—and Klaus Kinski's incandescent performance as Aguirre—brilliantly captures the madness of not just the protagonist but of the Conquest as a whole. The screenplay merges accounts of two major expeditions—that of Aguirre/Ursua and the earlier trek of Gonzalo Pizarro/Orellana, the first documented European voyage down the Amazon—in order to create a richer story.

Herzog made no secret of this: a major character in the film is Gaspar de Carvajal, the Dominican friar who accompanied Orellana and whose detailed diary serves as the basis for what we know of Orellana's journey. Some of the more unforgettable scenes and sequences are derived from other historical events occurring during the Conquest: the Yagua Indian who is handed a Bible and discards it because it does not speak to him is based on the response of the Incan emperor Atahualpa prior to the Battle of Cajamarca, and the conquistadors' vision of a European ship perched in a treetop is drawn from an account of the explorer Alvar Nunez Cabeza de Vaca, who observed this in Hispaniola after a hurricane—meteorologically common in the Gulf of Mexico but nonexistent in the Amazon. Herzog utilizes these episodes and accounts to weave together a haunting film that matchlessly captures the greed, stupidity, homicidal behavior, and irredeemable evil that characterized much of the Conquest.

What role did Pedro Teixeira play in the colonization of the Amazon?

Once the conquistadors had vanquished the great indigenous civilizations of Central America and the Andes, these lands were held firmly in the Spanish grasp. On the other hand, the British, Dutch, and French had established beachheads in the northeastern shoulder of South America—the region known today as the Guianas—as well as on islands in the Caribbean.

In what is now Brazil, the Portuguese—who considered themselves the rightful "owners" (based on the Treaty of Tordesillas) of these regions—were infuriated by the presence of Dutch, English, French, and Irish colonies, forts, and trading posts. That the larger and richer Spanish Empire ruled the Andes to the west—unchallenged by other European powers—undoubtedly gave the smaller and poorer Portuguese the equivalent of a colonial inferiority complex.

No Portuguese soldier played a more active role than Captain Pedro Teixeira. Born in Portugal in the late 15th century, he shipped out to Brazil to serve the Crown and personally and successfully fought against the Dutch, English, and French, helping to evict them from his adopted country. And it was to this renowned warrior that his countrymen turned when the arrival of a canoe from the west bearing eight Spaniards paddled into Belém harbor at the mouth of the Amazon. This canoe shocked and frightened the Portuguese, who assumed that the Spaniards were moving east into the Amazon.

Their worries that this little boatload of Spaniards was the advance guard of an invasion force proved unfounded. The canoe contained two Franciscan friars and six soldiers—and nobody was behind them. The Franciscan order had established a mission on the Rio Napo in what is now Ecuador and earned the ire of the local tribespeople, who had chased them out. Most of the Franciscans headed west to Quito, but these two had decided that God wanted them to travel downstream.

In 1637, an expedition under Teixeira's leadership was quickly organized and dispatched upriver in the ongoing continental rivalry between Portugal and Spain to determine who controlled the Amazon. Teixeira's band was the first of the Europeans to *ascend* the Amazon: Orellana and Aguirre had traveled downstream. Of course, Teixeira accomplished this feat powered by others: in this case, 47 canoes, paddled and rowed by Amerindians and Afro-Brazilians—it was unlikely that Teixeira did any of the heavy lifting. Noteworthy, though, is that he explored the mouth of the Rio Negro, the largest river emptying into the Amazon, and "discovered" and named the Madeira—the second largest. His secret mission was to place a stone boundary marker featuring the Portuguese coat of arms near where today Brazil, Colombia, and Peru meet, thus staking a claim thousands of miles to the west of the boundary established by the Treaty of Tordesillas.

Teixeira and his crew reached the first Spanish settlement 8 months after their departure and arrived in Quito 4 months later. As was the case with the arrival of the Spanish Franciscans at the mouth of the Amazon in Portuguese territory, the newcomers were feted as honored guests—but detained by local authorities, who did not know what to do with them. Finally, Teixeira was sent back home accompanied by Spanish Jesuits, the most noteworthy being Father Cristobal de Acuna, whose report was notably sympathetic to the Amerindian cultures they visited in the course of their downstream journey. The only sour note was the Portuguese desire to enslave tribal peoples along the way—a misadventure that was cancelled at Acuna's request.

Ultimately, the Teixeira expedition served as the opening gambit by the Portuguese to claim most of the Amazon Basin. The subsequent century of slaving and missionary work throughout Amazonia by the Portuguese resulted in—according to John Hemming—the Portuguese claim to much of the Amazon based on *uti posseditis* ("as you possess," meaning that the territory remains with its possessor at the end of a conflict) and this expansive claim was accepted by the Spanish at the negotiations leading up to the Treaty of Madrid in 1750, slightly more than a century after Teixeira's voyage.

What is the history of the mapping of the Amazon?

As noted earlier, some part of the conquest of Amazonia was cartographic. A most unlikely cartographer created the first maps of Amazonia: Samuel Fritz, a Jesuit priest born in Czechoslovakia in 1654. After studying geodesy and surveying, he was ordained at the age of 20 and sent to South America as a missionary in 1684. After a short residence in Quito, he was dispatched to the upper Amazon, where he began proselytizing from the upper Marañón in the southwest as far east as the Rio Negro, working in what is today Peru, Ecuador, Colombia, and Brazil, an astonishingly broad swath of Amazon rainforest territory to cover almost a century prior to the American Revolution.

Fritz possessed broad talents and training, as he was a carpenter, physician, skilled linguist, and student orchestra conductor. And—unsurprisingly—he was no stranger to hardship: his Amazon diaries recount an instance where he lay sick in his hammock in the throes of a malarial fever while his camp was flooded by a rising river and overrun by bellowing black caimans as rats devoured his provisions. Failing to improve, he traveled more than 3,000 miles (4,828 km) down the Amazon to the city of Belém in search of healing. This being the era in which the Spaniards and the Portuguese were tussling back and forth, east and west, to see who would control how much of the Amazon, he was arrested and imprisoned by the Portuguese as a Spanish spy despite his Czech origin.[3]

Fritz made optimal use of both the voyage and his year in confinement. Having likely been the first trained surveyor who not only lived in the Amazon but traveled from up near the headwaters down to the mouth at Belém, while captive, he produced an astonishingly accurate map of the entire river featuring both the names and locations of many of the tribes living along its banks. A half-century later, the renowned French geographer Charles Marie de La Condamine used Fritz's document as his basis for the map of the Amazon he produced after he had sailed the length of the great river once he had concluded his expedition in the highlands of Ecuador to measure the shape of the earth.

The Treaty of Madrid in 1750 brought an end to the armed conflict between the Portuguese and Spanish Empires in South America and established the basic frontiers between Brazil and the Spanish-speaking countries in Amazonia. This in turn ignited an explosion of efforts to establish borders in the most remote corners of the Amazon. The most prolific explorer in Brazil was Ricardo Franco de Almeida Serra, who led major mapping expeditions in both the northeast and southeast Amazon. Other prominent explorers who mapped the Brazilian Amazon were Alexandre Rodrigues Ferreira, Manoel da Gama Lobo d'Almada, Antonio Pires da Silva Pontes, and, the most famous of all, the heroic Candido Rondon. Rondon is best known for leading the famous Roosevelt–Rondon Scientific Expedition down the River of Doubt (now the Roosevelt River) in 1913–1914; he also laid the first telegraph wires across southern Amazonia, visited most of Brazil's remote border regions, and led an exploratory commission that published more than 100 volumes of scientific data and maps. Rondônia state in southwestern Brazil was named in his honor.

German brothers Richard and Robert Schomburgk carried out important explorations and mapping on both sides of the Brazil and British Guiana borders in the 1840s. The French couple Henri and Olga Coudreau also explored the northeast Amazon. Henri died of malaria in the rainforests of the Trombetas River; Olga continued the work for years afterward, making her one of the most experienced female explorers of the Amazon. Together, the couple published well over a dozen volumes of their studies of the rivers they explored.

The hideous Amazon rubber boom at the beginning of the 20th century resulted in numerous nonindigenous peoples flocking into the forests of the northwestern Amazon in search of "white gold." One of the very few and perhaps only positive consequences of this short-lived economic "miracle" was the mapping of the many rivers and forests that had been heretofore unexplored by the outside world.

The most colorful character in the history of Amazonian cartography was Dr. Alexander Hamilton Rice. Trained as a physician at Harvard Medical School, Rice fell in love with the rainforest while still a student and later decided to learn mapmaking skills by enrolling in classes at the Royal Geographic Society in London. His marriage to Eleanor Widener, one of the world's wealthiest women, turbocharged his career, and he was able to build planes that could map the Amazon from the air, a foretaste of the remote imaging technology that has now become commonplace.

The most important mapping of the Amazon in the latter half of the 20th century was carried out by the Instituto Brasileiro da Geografia e Estatistica (IBGE), which mapped all of Brazil. In 1970, IBGE launched Projeto RADAM, a 15-year effort to map the country using side-scanning radar to peer beneath the forest canopy, impossible from the air. It was this program, utilizing this approach, that found the world's largest iron ore deposit at Serra dos Carajás in the eastern Amazon.

In recent times, the Indians themselves have been undertaking mapping efforts aimed principally at better management and protection of their own lands. The Trio tribe in Suriname began working with the Amazon Conservation Team and other organizations to conduct one such effort in 2000, integrating their traditional territorial knowledge with geographic information system technology. Beginning with a few hand-held GPS units and some rudimentary government maps, the effort has continued to spread and improve, using extraordinarily detailed aerial maps and drones. To date, the Amazon Conservation Team program has partnered with 55 South American tribes and has mapped and improved the protection of 80 million acres of ancestral forests.

At the outset of the Trio project, aerial imagery consisted of pixels at 98 foot (30-meter) resolution; in 2017, using technology developed by DigitalGlobe, a pixel can resolve to less than 10 inches (25 cm). And new technologies like LIDAR—using lasers to map from the air—will continue to revolutionize mapping and our understanding of the Amazon. Given the rate of technological change, it will be possible to create maps depicting every tree in the rainforest, thus providing opportunities for better protection and improved stewardship of all Amazonian forests. But common sense and wise stewardship in Amazonia have seldom been the case since the advent of the Europeans.

Who was the first European scientific explorer of the Amazon?

Many date the beginning of the scientific exploration of the South American rainforest to Alexander von Humboldt and Aimé Bonpland's ascent of the Orinoco in February of 1800. Following their lead into the Neotropics were Carl von Martius and Johann von Spix (1817–1820) in the Brazilian Amazon; Charles Darwin in Brazil's Atlantic Forest (1832); Alfred Russel Wallace (1848–1852) and Henry Walter Bates (1848–1859) in the Brazilian Amazon; Richard Spruce in Brazil, Ecuador, and Peru (1849–1864); and Louis Agassiz in Brazil (1865–1866).

Missing from this august lineage is an almost-forgotten woman who worked, collected, observed, and illustrated more than a century before von Humboldt and Bonpland set foot on the South American continent. Unlike von Humboldt, Maria Sibylla Merian possessed no great personal fortune nor had she received any formal university education. In fact, at the time she set sail for the rainforest, she could have been accurately described as a middle-aged, unemployed, and divorced German housewife. What she shared with all the great explorer naturalists who followed in her footsteps into the Amazon rainforest was a passion to immerse herself in the unknown and to study and document the creatures that dwelled within.

Merian was born into a family of artists and printers in Frankfurt in 1647. She became entranced with nature at an early age: first with flowers, and then with insects, particularly caterpillars and moths, beginning with silkworms. At a time when many believed that insects arose spontaneously from plants or rotten meat, she proved by lengthy and detailed observations that caterpillars metamorphosed into butterflies and moths, publishing a richly illustrated book on the subject at the age of 22.

Following an unhappy marriage, she moved to Amsterdam in 1691. There she earned a living selling botanical paintings, taking advantage of the Dutch passion for both art and flowers. At the time, the Dutch Empire was flourishing, and Amsterdam was a global crossroads. Bizarre and beautiful plant and animal specimens poured in from both the East and West Indies. The dazzling colors, shapes, and sizes of these tropical organisms inspired her to visit, observe, collect, and paint them in their native habitat. On July 10, 1699, she set sail for Paramaribo, the capital of Dutch Guiana, today the independent nation of Suriname.

Although north of the Amazon Basin proper, Suriname, British Guiana (now Guyana), and French Guiana are part of the same great rainforest ecosystem. And, unlike Brasilia, Bogotá, Caracas, La Paz, Lima, and Quito, the capital cities of the Guianas (Cayenne, Georgetown, and Paramaribo) are located within the Amazon rainforest, meaning that Maria Sibylla Merian in Paramaribo was well situated to collect, study, and illustrate insects and plants seldom seen by the outside world. Not only were her paintings of these species exceedingly accurate, but the interactions between the plants and the animals that she documented were revolutionary. Merian illustrated the ecological interactions and struggles among Amazonian organisms more than 150 years before Ernst Haeckel coined the term "ecology" and more than 160 years prior to Darwin's publication of his immortal work on the competition of species for survival.

So precise were her paintings that nearly three-quarters of the butterflies she depicted can be identified to the genera in which they are now classified, and more than 66% can be identified to the species level. The great Linnaeus himself considered her artwork so accurate that he used it as the basis for describing dozens of new species. Some of the biological interactions Merian illustrated were so unusual that they provoked ridicule among her peers: her depiction of a tarantula devouring a hummingbird was not deemed credible until entomologist Henry Walter Bates recorded a similar incident with two finches in Brazil almost 150 years later.

Merian's powers of observation were acute. She noted that some Amazonian ants build bridges with their bodies, others carry leaf pieces into underground burrows, and some can quickly strip vegetation—these were undoubtedly among the first accounts of leafcutter ants and army ants. Her observations on agronomy were similarly sophisticated. She produced some of the first paintings of cacao, cashew, cassava, and pineapple, and she unsuccessfully attempted to convince local planters to cultivate these species to diversify beyond their sugarcane monocultures.

Her actual itinerary in Suriname is no longer extant, but we do know she traveled and collected as far south as La Providence, 30 miles (48 km) up the Suriname River south of Paramaribo. Her collecting efforts were aided by local Amerindians and enslaved Africans who were undoubtedly fascinated by the novelty of a European woman who had ventured into their rainforest and wanted to learn from them. She in turn demonstrated her respect for these people and their knowledge, recording their names and their uses for local species, thereby carrying out some of the very first ethnobiological research in the Amazon.

These interactions with local colleagues brought forward another aspect of Merian's personality little noted in most biographical sketches: her humanism. She fearlessly condemned the Dutch treatment of local slaves. Foreshadowing Charles Darwin—who would write movingly of an Afro-Brazilian woman who committed suicide rather than face a life of slavery—Merian wrote,[4]

> The Indians, who are not treated well by their Dutch masters, use the seeds [of *Caesalpinia pulcherrima*, the so-called peacock flower or pauwenbloem] to abort their children, so they will not become slaves like themselves. The black slaves from Guinea and Angola have demanded to be

well-treated, threatening to refuse to have children. In fact, they sometimes take their own lives because they are treated so poorly, and because they believe they will be born again, free and living in their own land. This they told me themselves.

What is the connection between the Amazon and the origins of the theory of evolution?

When asked about the origin of the theory of evolution, most attribute it to Darwin's famed visit to the Galápagos Islands in late 1835. While this was a seminal event in the development of evolutionary biology, the story did not begin and end with the British biologist's visit to this Pacific Ocean archipelago.

Darwin's voyage aboard the HMS *Beagle* commenced during the final days of 1831 and lasted almost 5 years. Though it was a global surveying expedition, 4 of the 5 years were spent in and around South America. While the most famous segment of Darwin's trek was his sojourn in the Galápagos (the islands today are considered part of Ecuador), he also visited Argentina, Brazil, Chile, Peru, and the Falklands, providing him with the broadest first-hand perspective on the natural history of South America of any biologist, then or since.

Though Darwin never ventured into the Amazon, he did explore the Atlantic Forest in both Rio de Janeiro and Salvador da Bahia—where the forest is very similar in structure and appearance to Amazonian rainforest—and was stunned by its beauty and the diversity. "Delight," he recorded in his journal, "is a weak term to express the feelings of a naturalist who for the first time has wandered by himself in a Brazilian forest . . . such a day brings a deeper pleasure than he can ever hope to experience again."

Darwin's rapturous experience in the South American rainforest, his subsequent encounter with the dwindling population of the Falklands Island wolf, and his discovery of the fossils of extinct titanic creatures in Patagonia—all before he had ever seen the Galápagos—started him thinking about the origin and extinction of species long before he first set foot in those remote islands.

Prior to his South American travels, Darwin had been prepared and trained for this experience by a man born in Amazonia. In 1825, 6 years prior to embarking aboard the *Beagle*, Darwin enrolled in medical school at the University of Edinburgh. However, unable to tolerate either the sight of blood or surgical operations—this being prior to the use of anesthesia—he soon realized he would make a poor physician.

The young Englishman decided to take lessons in taxidermy from John Edmonstone, a freed slave who lived a few doors down from him on Lothian Street in Edinburgh. While Edmonstone taught the budding biologist and natural history enthusiast how to stuff animals—an invaluable skill during his epic voyage—he undoubtedly regaled him with stories about the fabulous birds and other creatures of the Amazon rainforest. Edmonstone had been born and raised in British Guiana, where he had learned the art of taxidermy from the swashbuckling British explorer Charles Waterton. In fact, Edmonstone likely served as Waterton's local guide on one of his treks through more than 400 miles (644 km) of the uncharted rainforests of the northeast Amazon.[5,6]

Meanwhile, 12 years after Darwin returned to England from the voyage of the *Beagle*, Alfred Russel Wallace, then an underemployed British surveyor, set sail for the Brazilian Amazon, intending to experience the rainforest and at the same time earn an income by collecting exotic insects and sending them to England for sale. Wallace was accompanied by Henry Walter Bates and, a year later, by botanist Richard Spruce. While the three eventually went their separate ways in South America, they maintained a lifelong friendship and collaboration. And Wallace's thinking on the origin of species was greatly influenced not only by his own studies in the Amazon rainforest but by Bates's research as well. For example, Bates repeatedly observed how the coloration of unpalatable insects was mimicked by more tasty species for protective purposes, a process now known as *Batesian mimicry* and convincingly explained only through the prism of evolution.

Wallace spent 4 years in the Amazon rainforest and later lived, worked, and collected in the Malay Archipelago for more than a decade. He always stated that the theory of evolution came to him in a hammock in the Maluku Islands during a malarial dream. Wallace sent his hypothesis on the evolution of species in a letter to Darwin, who had long been thinking along the same lines, and they published this together in 1858.

Thus, the theory of evolution was not Darwin's alone, nor did it begin solely in the Galápagos. The Amazon forest and the life and experiences of an ex-slave played a vital, if little-known, role. In recognition of his importance, John Edmonstone was recently named to the BBC's list of "100 Great Black Britons."

Did Harvard send an expedition to the Amazon to disprove Darwin's theory of evolution?

Widely regarded as one of the preeminent biologists of the 19th century, Louis Agassiz served as a founding father of American science. Born in Switzerland in 1807 as Jean Louis Rodolphe Agassiz, he trained in Germany as both a physician and a botanist, a common practice at the time. He then moved to Paris to study fossils under the tutelage of the great French paleontologist Georges Cuvier. There, he fell under the sway of the prominent German naturalist Alexander von Humboldt, who was already widely renowned for his explorations in South America and who undoubtedly fired Agassiz's imagination with tales of the great rainforest.

At this point in his life—even though he had never set foot outside of Europe—Agassiz was already engaged with the Amazon. The Portuguese who ruled Brazil had repeatedly fended off colonization attempts by the British, English, French, and Spanish, as well as by lesser powers like the Irish(!), and so were exceedingly paranoid about foreigners setting foot in their colony. As a result—in the words of historian John Hemming—"all Brazil and the Amazon were rigorously closed to foreigners."[7]

In 1808, the Portuguese court abandoned Lisbon for Brazil to escape Napoleon's advancing army. King João ruled the Portuguese empire from Rio de Janeiro until he returned to Europe in 1821, and his heir Dom Pedro stayed behind and proclaimed himself emperor of Brazil the following year. In 1817, he had married the Austrian archduchess Leopoldine ("Leopoldina") Caroline. At the instigation of the far-sighted Austrian chancellor von Metternich, she had come to Brazil accompanied by a team of scientists intent on conducting research and (in all probability) looking for new resources. Two of the scientists—botanist Carl von Martius and zoologist Johann von Spix—traveled and studied in Amazonia in 1819 and 1820 and were the first non-Portuguese Europeans to obtain official permission to conduct research in Brazilian Amazonia. Researchers study their collections of plants and animals—still housed mostly in German museums—even today.

Martius lived a long and productive life, but Spix sickened and perished shortly after returning to Europe. In 1827, Agassiz was studying at the University of Munich. Martius—then one of his professors—asked his brilliant student to study and scientifically describe the Brazilian fish (many from the Amazon) collected by Spix, the first volume of which the young Swiss biologist completed and published in 1829 and the second 2 years later. At this point, Agassiz was 24 years old.

Because of his knowledge of the extant Brazilian fish fauna, he chose to focus his studies on Brazilian fish fossils while working under Cuvier at the Jardin des Plantes. The result was a multivolume work, the final book being published in 1835. At this point, Agassiz was the world's leading authority on both fossil and living fishes in Brazil despite never having set foot in that country.

Meanwhile, in Brazil, Dom Pedro II has ascended to the throne. He proved to be a far-sighted and progressive leader, promoting civil rights and abolishing slavery. He was fond of the Indians—not a common sentiment among South American leaders in the 19th century—and had an abiding passion for natural history. Following in his mother's footsteps, he was supportive of scientific studies of the Amazon, even if it was carried out (in part, at least) by foreign scientists.

In 1846, Agassiz traveled to the United States to present a series of lectures and—in a professional sense—never left. So dazzling were his discourses in Boston that he was offered a professorship at Harvard. Through willpower, creativity, fundraising, and fanatical dedication to building a scientific establishment in his adopted land, Agassiz founded Harvard's Museum of Comparative Zoology, one of the first publicly funded scientific buildings in North America and a leading natural history institution to this day. The indefatigable Agassiz also found time to teach, to urge the creation of the National Academy of Sciences, and to found the precursor to the esteemed Woods Hole Oceanographic Institute, all the while building collections for his new museum at Harvard. This is in addition to his pioneering work on glaciers and geology: Agassiz was among the first scientists to propose that much of the earth had once been covered by sheets of ice.

Agassiz clearly exulted in the attention and adulation and was widely considered to be the most famous scientist in his adopted country. But two flaws have stained much of his legacy: over the course of time, he proved to be both a racist (believing that different races were essentially different species) and a creationist.

When Darwin published *On the Origin of Species* in 1859, Agassiz was at the height of his fame. He was widely celebrated for his brilliance, his eloquence, and his ability to perceive things beyond the discernment of mere mortals. He was also arrogant, bombastic, and stubborn. Agassiz believed in the immutability of species—that species could not and did not evolve over time. Unfortunately, his closed-mindedness and unwillingness to reconsider Darwin's brilliant theory proved immutable as well.

At Harvard, the reputation of the brilliant, beloved, and seemingly infallible Agassiz was being eclipsed by the more mild-mannered botanist Asa Gray, whose defense of Darwin proved compelling, effective, and correct.

Mulling a masterstroke that could disprove Darwin while winning back the acclaim he considered his due, Agassiz conceived of an expedition to far-off Brazil, the unseen land that had first fascinated him more than three decades earlier. Agassiz believed that the Thayer Expedition—named for the Harvard benefactor who underwrote the costs—would allow him to find and name species of fish then unknown to science, augment his new museum's burgeoning natural history collections, study patterns of fish distribution, and find evidence of local glaciation with which he could disprove Darwin. The expedition party sailed out of New York harbor on the steamer *Colorado* on April 1, 1865, slightly more than a week before the cessation of the US Civil War.

The expedition officially commenced in Rio de Janeiro, where Emperor Dom Pedro, delighted to have such an internationally famous scientist studying in Brazil, personally welcomed the party and offered them all available assistance. The team spent 3 months collecting on and near the Atlantic coast of Brazil before sailing into and up the Amazon River.

Greatly aided by Major Joao Martins da Silva Coutinho, a geologist and naturalist of the Brazilian Army Corps of Engineers who had been seconded to them by the emperor, they collected fish, plants, and other materials over an astonishing range, branching out from the central stem of the Amazon River to also gather specimens from the Rio Branco, Rio Ica, Rio Javari, Rio Negro, Rio Madeira, Rio Tapajos, Rio Tocantins, and Rio Xingu. In fact, the research party reached Tabatinga, a little town that sits on the Brazil–Colombia border, thousands of miles west of the Amazon estuary. The Thayer Expedition's Amazonian foray took most of a year, and they set return sail for the States on July 2, 1866.

Even today, more than 150 years later, there is still no accurate tally as to precisely how many specimens were collected, nor have all of the specimens been scientifically identified. Agassiz was never able to prove that "the hand of God" was responsible for the way in which fish were distributed in Amazonia, nor did he find evidence of glaciation that would refute Darwin's theory. He and his team did make remarkable collections that remain cornerstones in our understanding of tropical South American fish fauna. And two other unforeseen circumstances did in some way manifest at least as a part of this Amazonian voyage. Louis's wife—Elizabeth Cary Agassiz, a brilliant educator in her own right who accompanied Agassiz on this voyage—wrote a wonderful chronicle of the expedition, *A Journey in Brazil*. This tome helped her step a bit out from her husband's enormous shadow, and she eventually served as the first president of Radcliffe College.

One enduring legacy of the Thayer Expedition was the subsequent work of Charles Frederick Hartt. He was a student assistant to Agassiz at the Museum of Comparative Zoology when he was invited to be part of the expedition. He quickly became enamored with Brazil and went on to become a pioneer in Amazonian geology, archaeology, and petroglyph carvings. Hartt authored one of the first ethnographic overviews of Amazonia and became the founder of the Geology Department at the National Museum of Rio de Janeiro, the Brazilian equivalent of the Smithsonian, this despite perishing from yellow fever at the age of 38.

Another member of the Thayer Expedition was a Harvard Medical School student and Agassiz acolyte named William James. Up until this voyage, James had lived the sheltered life of an upper-class Bostonian. In Brazil, however, he lived, talked, traveled, and collected with the local gentry, ranchers, fisherman, farmers, peasants, Afro-Brazilians, and forest Indians of many different tribes. James returned home a much more enlightened scholar and completed his medical studies. He also became a philosopher and is now widely regarded as the father of American psychology. The Thayer expedition—and the Amazon itself—inarguably offered stimulus for James's comprehension of the human mind and condition far beyond that tendered by the rarefied atmosphere of American and European academia.

How did Amazonian rubber become a key global commodity?

In an age dominated by synthetics, several billion-dollar industries of the modern world still run—literally—on rubber, from the National Basketball Association (shoes) to the National Association for Stock Car Auto Racing (NASCAR; tires) to the airline industry (ditto). Only natural rubber, which originated in the Amazon, has the needed durability and elasticity.

An observation that foreshadowed the development of high-end athletic shoes was made in the northeast Amazon by the French ethnobotanist Jean Baptiste Christophore Fusée Aublet in 1763, when he observed local Indians dipping their feet into *Hevea* rubber tree sap and holding them over the fire, thus creating the first bespoke cross-training shoes.[8]

Amazonian rubber was originally brought to Europe in 1745 by the French geographer Charles Marie de La Condamine, who shared the peculiar substance with scientific colleagues in Paris upon his return from South America. La Condamine was said to have employed the latex to create waterproof bags to keep his scientific instruments dry during his South American sojourn.[9]

Another European, the Glaswegian Charles Mackintosh, increased commercial demand for the substance when, in 1823, he combined it with naphtha, converting rubber into a pliable and waterproof coating for fabric. The eponymous raincoat known (in the United Kingdom) as the mackintosh is named after him. Though Mackintosh's process and creation represented an important step forward, the rubber still tended to get sticky during hot summers and stiff and brittle in cold winters. An American, Charles Goodyear, augmented the utility of rubber further still in 1839, when he added sulfur in the presence of heat, thereby significantly improving the mechanical properties of the Amazonian latex and making it more resistant to heat and cold.

With the upgrading of rubber's mechanical properties due to Goodyear's process, the industrialized world now had a soft, springy, waterproof substance that retained its shape when bent and retained these properties across temperature ranges.[10]

Meanwhile, the Industrial Revolution was well under way, and the world was transitioning from water to steam power. Steam was moved through steel pipes, and gaskets made of natural rubber provided the best seal between connecting pipes. And with the invention of the pneumatic tire and the first mass production of automobiles in 1901, the stage was set for demand to soar. At this time, the Amazon remained the sole source of high-grade rubber.

As natural rubber changed from a scientific curiosity to a vital element of new technologies, the Amazon rainforests soon teemed with people in search of what was termed "white gold." Titanic fortunes were amassed and obscene crimes were committed in pursuit of those fortunes, culminating in the Putumayo Atrocities.

Today, almost all natural rubber comes from Amazonian trees planted in southeast Asia. It still offers an elasticity, plasticity, and resistance to abrasion that synthetic rubber cannot duplicate. In addition to the aforementioned car and airplane tires, natural latex is employed in the manufacture of airbags, wetsuits, adhesives and protective coatings, surgical gloves, condoms, and automobile parts such as serpentine belts and brake padding. In the words of historian Charles Mann, "Modern machinery uses natural rubber much more often than it did in the past because it is typically working at levels of higher stress and precision."

What effect did the rubber industry have on Amazonia?

Throughout the early 19th century, rubber was exported from the Amazon rainforest almost as a novelty item, used mainly for pencil erasers, toys, garters, and other articles of limited commercial value. After the invention of vulcanization in 1839, which markedly improved its mechanical properties and augmented its resistance to temperature variations and its versatility, however, the demand for rubber exploded. Prior to this, bicycles rode on hard and uncomfortable tires made of wood, metal, and solid rubber. The new and improved rubber made possible the creation of pneumatic (air-filled) tires and demand further increased. Rubber was also used to coat everything from underwater cables to zeppelins. The founding of the automobile industry created a demand for millions of rubber tires. In 1855, somewhere around 2,000 tons of rubber were being exported from the Amazon. By 1870, exports increased fivefold.

Rubber came from the Amazon rainforest, but there had been little or no effort to create plantations of rubber trees: after all, demand was limited and the rainforest seemed limitless. Even as demand spiked sharply in the 1860s, few seemed interested in or patient enough to plant rubber seeds and wait years for the trees to produce latex. And debt peonage systems were already in place whereby forest residents would buy overpriced manufactured goods from traveling traders in return for agreeing to collect forest products like medicinal plants or animal pelts.[11] Now the demand shifted to rubber. Most rubber merchants—personified by the unscrupulous Peruvian Julio César Arana—saw that creating their own paramilitary force, claiming large tracts of rainforest as their private fiefdoms, and enslaving the local Indians would prove very lucrative.

The Putumayo River Valley in the northwest Amazon was rich in rubber and home to thousands of Witoto and Bora Indians and other tribes. Because the national borders of Colombia, Ecuador, and Peru had not yet been officially established, Pope Pius X declared in 1906 that all armed forces must depart the area until the boundary questions were settled, thus creating a vacuum that Arana and his murderous thugs quickly filled.

As rubber poured out of the rainforest, money poured into the Amazon. Sleepy towns like Manaus and Belém became important commercial centers, featuring department stores, cobblestones shipped in from Portugal, a racetrack, a bullring, a world-class opera house, and Brazil's first telephone system. Rubber barons competed to see who could build the biggest and most ornate palatial residences while shipping both their children (for education) and their soiled clothing (for laundering) to Europe.

According to Richard Collier, who vividly chronicled this Amazonian age of excess in *The River That God Forgot*, the diamond trade boomed during the rubber boom, as rubber barons gifted these gems to their daughters, wives, and mistresses. They wore the finest tailored clothes, lit cigars with large bank notes, and watered their horses with French champagne. The per capita income of Manaus—essentially a jungle outpost—was twice that of thriving Brazilian coffee ports like Rio de Janeiro and São Paulo.

Others paid the price. Local Indians were enslaved, tortured, raped, mutilated, and often murdered in the quest to obtain ever more rubber, a hideous episode that became known as the *Putumayo Atrocities*. With no oversight, Arana and his henchman ran one of the most vicious regimes in recorded history, though at the time few in the outside world were aware of what was happening. It is estimated that as much as 80% of the local population was exterminated in the drive to meet the industrialized world's growing demand for rubber.[12] Only the brave and pioneering work of American journalist Walter Hardenberg and British diplomat Sir Richard Casement finally brought these crimes to international attention, deeply damaging the trade's reputation.[13] The other nail in the coffin of this rubber boom was the development of

Southeast Asian rubber plantations, whose lower prices undercut those of Amazonia.

Some Brazilian nationalists still complain about British businessman Henry Wickham stealing their national patrimony well over a century after he shipped rubber seeds from the Amazon to London to establish rubber tree plantations in British colonies in Asia, thereby breaking the South American monopoly on rubber production. What is undeniable, however, is that the establishment of these plantations in tropical Asia, which produced rubber more cheaply than that gathered from the wild in Amazonia, saved the lives of many indigenous peoples of Amazonia.

Many Colombians—whose nation endured some of the worst excesses during the Putumayo Atrocities—have a more fatalistic view of what happened with the demise of the local rubber trade, which they sum up in the aphorism *muerto el perro, se acaba la rabia!*: "When the dog died, the rabies died with him!"

What was Fordlandia?

Fordlandia was the brainchild of Henry Ford, who, like many exceedingly rich and powerful people, thought he knew and could control much more than he actually did and could. Ford was the founder and driving force behind the automobile industry, and that meant that he had created an enormous market for rubber tires.

By the 1920s, Southeast Asia was producing so much latex that a British Secretary of State for the Colonies (also known as the "Colonial Secretary") by the name of Winston Churchill had proposed the creation of a cartel to control the price of rubber by limiting production. Unlike the British and the Dutch, the Americans owned no rubber-producing colonies and were outraged. Ford, then building and selling about half the world's automobiles, had an almost insatiable appetite for rubber, from which he made the tires for his

Model T's and Model A's. He decided to act. He decided to grow his own.

Rather than hire experienced tropical foresters, soil scientists, ecologists, sociologists, or anthropologists at the outset, any one of whom could or would have saved him enormous amounts of money and heartache, Ford plunged ahead without the benefit of relevant expertise. He saw this project as not only a business enterprise but a chance to build an American Midwestern civilization in the middle of the South American rainforest. For him, both the local ecosystem and the local populace were tabulae rasae on which he could decide what needed to be inscribed.

He built a model town—christened "Fordlandia," though Ford himself never set foot there—featuring indoor plumbing, window screens, fire hydrants, a hospital, tennis courts, an 18-hole golf course, swimming pools, Baptist churches, sidewalks, and electric street lights, all manifestations of an American lifestyle. He offered better wages, sanitation, education, and healthcare than these Brazilians had ever seen.

The tradeoffs soon became apparent to local workers, however. He banned consumption of cachaça, the national liquor of Brazil, which is made from sugarcane (rum—with which it is often mistaken—is made not from sugarcane but from molasses) and which is the main ingredient of the caipirinha, Brazil's national cocktail. Though the Brazilians in Amazonia had long thrived on a diet of fresh fish, tropical fruits, beans, and rice, Ford insisted they switch to a diet of brown bread, brown rice, oatmeal, and canned peaches, resulting in a riot in 1930 that was only quelled by the arrival of the Brazilian military.

Being told what they should and should not eat and drink was not the only issue that infuriated the locals. Employees were expected to be grateful for the clapboard bungalows in which they now lived, but these dwellings were designed in and for the climate of Michigan and proved suffocating in the stifling Amazonian heat. Modeled on the American schedule,

Ford's employees were supposed to work from 6 AM to 3 PM, which had them toiling away during the hottest part of the day.

And Ford tried to impose his ascetic and puritan ways on the Amazon rainforest: workers were expected to refrain from engaging in traditional forms of dance, which were deemed too sensuous, instead learning more chaste American and European numbers, such as square dances, waltzes, polkas, and quadrilles. Rather than abstain, with typical Brazilian aplomb locals chose an island upstream, which they whimsically christened "Isla da Innocencia"—the Isle of Innocence—and there opened bars, bordellos, and gambling dens.

Ultimately, however, Ford's arrogance and his ignorance of local culture were perhaps less damaging to his project than his prejudice against expertise and his ignorance of rubber tree biology. For the first 5 years of the project, there was no one on the Fordlandia payroll who understood tropical agriculture, forestry, soil science, or rubber trees in particular. As such, they had launched their project on a site that was too sandy and too dry for optimal forestry production. Nor had they searched for or planted rubber trees that were high yielding and/or relatively pest resistant. The latter oversight ultimately doomed Ford's entire effort, for the rubber tree plantations were soon invaded by South American leaf blight (SALB), a fungus native to Amazonia that attacks, weakens, and usually kills the trees. The fungus that causes the leaf blight—*Microcyclus* (formerly known as *Dothidella*) *ulei*—lives in the Amazon rainforest. The spores cannot travel far, which is why individual rubber trees are not found growing closely together but are widely dispersed in the forest. Planting trees in close proximity to one another in the presence of the fungus is an ecological recipe for disaster. In fact, most plantation crops are cultivated far from their center of origin to evade the pests that coevolved with them.

In 1934, Ford's people recommended purchasing land downriver from Fordlandia that was closer to the mouth of the Tapajos River, hoping that better soils and more breezes would yield more rubber and reduce the threat of leaf blight. Known as Belterra, initial results at this site seemed promising—and then the dreaded SALB swept through there as well.

When Ford attempted to bend the Amazonian ecosystem and people to his will, he suffered the greatest business failure of his long life. In purely commercial terms, in 1927, he very likely spent close to $20 million (around $300 million in today's dollars) for 2.5 million acres of rainforest along the Tapajos River. Eighteen years later, his company sold his deforested land back to the Brazilian government for about $500,000.

What was the Jari Project, and why did it fail?

The Jari Project of American Daniel Ludwig thoroughly reinforces George Santayana's famous aphorism about repeating the unremembered past—in this case, with reference to Henry Ford's Fordlandia misadventure.

Like Ford before him, Daniel Ludwig was born in rural Michigan and never earned a college degree. In the 1930s, Ludwig began building what would become one of the world's largest shipping companies, National Bulk Carriers, one of the first enterprises to manufacture and maintain a fleet of oil supertankers. He then diversified into agriculture, banking, cattle ranching, insurance, mining, and tourism, with holdings in 50 countries. In the 1960s, Ludwig saw another opportunity: worldwide demand for paper and fiber was growing. Ludwig reasoned that he could make his next fortune in the Amazon, with its cheap land and labor and year-round growing season.

In 1965, he purchased an enormous concession from the Brazilian government—about 4 million acres along the banks of Brazil's lower Jari River in northeast Amazonia, roughly the size of the state of Connecticut. He chose to plant *Gmelina arborea*, a fast-growing tree from Southeast Asia that produces high-quality wood and leaves that can be eaten by cattle. Unfortunately, *Gmelina* does not thrive on the sandy soil that characterizes the Jari region. Moreover, the trees planted there were devoured by local pests and insects.

Unlike Ford, Ludwig did hire trained foresters to oversee operations, but these were temperate-zone foresters whose knowledge and skills proved inadequate in the tropics. Like Ford, Ludwig built a model town—Monte Dourado—equipped with paved roads, a railroad, schools, nurseries, bridges, community centers, and a hospital. Local workers were offered better wages, sanitation, education, and healthcare than they could find elsewhere in most of Amazonia. Unfortunately, Ludwig insisted on some of the same behavioral restrictions that had been placed on the workers in Fordlandia. And—as in Fordlandia—locals built a nearby town, Beiradao, filled with bars and brothels. Jari is perhaps best known for the 17-story pulp mill that Ludwig had constructed in Japan—where many of his ships were built—and shipped across the Indian Ocean, around the Cape of Good Hope, and then up the Amazon River.

Because the bulldozers employed to clear the rainforest had destroyed the thin layer of topsoil, the *Gmelina* never grew as well or as quickly as had been projected, meaning that Ludwig began experimenting with other tree species like Caribbean pine (*Pinus caribaea*), which also failed to yield the rate of pulp production projected. Ludwig also ventured into cattle ranching and rice farming, but he never turned a profit.

Like Ford before him, Ludwig alienated local sensibilities. His penchant for secrecy—reporters were kept out of Jari for more than a decade—and his preference for non-Brazilian senior staff fueled local xenophobia. Headlines in the local press such as "The Jari Project—The American Invasion," a book, *Ludwig: Emperor of Jari*, and a play, *Jari: The Country of Mister Ludwig*, further stoked Brazilian paranoia. In 1981, an aged, angry, and ill Ludwig pulled out of Jari, having lost an estimated $700 million on the venture.

Has Amazonia produced any heroes?

Yes, it has, and Cândido Rondon and Chico Mendes should be counted among them.

Rondon was born in Mato Grosso in southern Brazil in 1865. His father was of Portuguese descent and his mother was of mixed ancestry, including at least some Bororo tribal heritage. Rondon joined the military at the age of 18 and soon proved his mettle as an army engineer, laying telegraph lines through much of southern Brazil and into Bolivia and Peru. At the time, this type of effort not only entailed working in and mapping previously uncharted territories but also gaining the support (or, at least, the acquiescence) of tribes known to kill outsiders who trespassed on their lands. Rondon is justly feted for having laid thousands of miles of telegraph wire through much of the Amazon, thereby knitting together much of the country. He surveyed many of Brazil's borders with other countries, and led the 1914 Roosevelt–Rondon expedition down the River of Doubt.

Rondon has long been considered a Brazilian national hero, one of the earliest who celebrated his mixed-race heritage. Historian John Hemming noted that Rondon was one of the first Brazilians to ascend to "rock star" eminence: admirers covered him with garlands of flowers when he returned from the bush, and his lectures were sold out. More importantly, he used his fame to champion the indigenous cause with regard to protection of both their cultures and their lands. Rondon was an effective spokesman against forced religious conversion and played a leading role in the establishment of the Brazilian government's Indian Protection Service (originally known as the *Servico de Protecao aos Indios* [SPI]; now called *Fundacao Nacional do Indio* or the National Foundation of the Indian [FUNAI]).

Rondon was fearless in making first contact with hostile tribes—at his encouragement, the motto of FUNAI became "Die if you must, but never kill!" ("Morrer se preciso for . . . matar nunca!"). In his early career, he believed that indigenous peoples should be acculturated and eventually absorbed into Brazilian society, but later he came to believe that these people should have the right to live a life and follow a culture of their own choosing on their traditional lands. He is still widely revered in Brazil: thousands of streets, schools, and even a state (Rondônia) are named in his honor.

Francisco Alves Mendes Filho—more widely known as Chico Mendes—was born in a rubber extraction reserve outside the little town on Xapuri, in the state of Acre in southwestern Brazil, in 1944. Forty years later, roadbuilding had brought outsiders and developers into the region, resulting in rapidly accelerating deforestation by people who wanted to fell the forest for timber and/or create cattle ranches. These developers were opposed by many locals who preferred to protect the rainforest and manage it sustainably, generating income by tapping rubber or extracting other non-timber products such as Brazil nuts (*Bertholettia excelsa*).

From among the ranks of these rubber tappers rose Mendes, by this point a trade union leader who did not learn to read until he was 18 years old. Finding he had a talent for organizing and public speaking, Mendes became an ambassador for the environmental movement, speaking out about debt peonage, the inequities of the Amazon economy, and the need to protect and care for the forest. When he gained the ear of multilateral organizations, including the Inter-American Development Bank and the World Bank (institutions then decried by environmentalists for the destruction unleashed by their infrastructure development projects), Mendes earned the ire of powerful Brazilian commercial interests. He was murdered by local ranchers near Xapuri in 1988.

Today, his legacy endures in the form of the Chico Mendes Extractive Reserve, which was created to honor him following his assassination; other extractive reserves were launched along the same lines. The Chico Mendes Institute for the Conservation of Biodiversity is run by the Brazilian Ministry of the Environment. The struggle continues: more than 30 years after Mendes's murder, more environmental activists are killed in Brazil each year than in any other Amazonian country.

7

AMAZONIA'S UNCERTAIN FUTURE

I have already outlined a number of threats to Amazonia like cattle ranching, mining, and climate change, but what follows are some of the most urgent—and ominous—risks and challenges that lie ahead.

Is cattle ranching a major cause of Amazonian deforestation?

Brazil possesses one of the world's largest cattle herds, estimated at over 200 million head. However, cattle ranching in South America began not in Brazil but in Colombia, in 1525. By the early 1600s, the Jesuits had established missions far to the south in Paraguay, in which they were raising cattle. There, members of the local indigenous Guarani tribe served as ranch hands, meaning that some of the very first American cowboys were in fact Indians.

The Jesuits were expelled from South America in 1767, but cattle ranching began to take root with local populations from the pampa grasslands of central Argentina through Uruguay into southernmost Brazil, inaugurating the famous gaucho culture that still survives in that region. Meanwhile, there were desultory efforts to raise cattle in Brazil's parched northeast, but the lack of water generally quashed this enterprise.

Until relatively recently, the lack of both pasture and roads and (to a lesser extent) the presence of jaguars and vampire bats largely discouraged cattle ranching in Amazonian rainforests. The major exception took place at the turn of the past century, during the Amazon rubber boom. Rapid population growth and the huge fortunes being generated in Manaus fueled a demand for beef. The Makushi (the same people who had taught the use of curare to Charles Waterton in the early 18th century) and the Wapishana tribes living on both sides of the Brazil–Guyana border began raising cattle on the savannas of Roraima and brought the animals to market on barges down the Rio Branco and Rio Negro.

Prior to the 1960s, most cattle ranching in the Brazilian Amazon focused on raising water buffalo (*Bubalus bubalis*) on Marajo Island at the mouth of the Amazon. Thereafter, the construction of roads in Amazonia facilitated access to markets and brought in Brazilians from other parts of the country who were knowledgeable about raising livestock. Zebu cattle well adapted to tropical climes were imported from India, and the government provided fiscal incentives with the stated goal of turning Brazil into a major beef exporter. Few realize that the Brahman, one of the first beef cattle bred in the United States and currently ranked among the world's most popular breeds, was developed from cattle imported into the United States from both Brazil and India.

In creating a major cattle industry, the Brazilian government and private sector exceeded even their most optimistic projections. Today, Brazil is one of the world's biggest beef exporters, shipping meat to 150 countries and supplying about a quarter of global demand. Estimates are that this industry generates more than $10 billion a year in export revenue.

About 40% of the Brazilian cattle are in Amazonia, and the astronomic growth of this industry has taken a serious toll on the rainforest. Approximately 70% of deforestation in the Brazilian Amazon is tied to beef production. Greenpeace has calculated that the cattle sector in Brazil is the single largest driver of deforestation in the world. And the government has announced plans to sharply increase cattle production.

Most of the beef produced in Brazil is consumed domestically. However, Brazil has (along with China) become one of the top producers of tanned leather, and the majority of this commodity is shipped overseas, where it is used to make shoes, furniture, vehicle upholstery, and high-quality goods for the fashion market.

In 2009, Greenpeace launched its famous (some Brazilian government officials prefer to think of it as infamous) "Slaughtering the Amazon" campaign, detailing how much forest destruction was being driven by the burgeoning cattle sector. Perhaps most devastating was Greenpeace's claim that major global brands like Nike, Adidas, Carrefour, Walmart, Kraft, Timberland, Louis Vuitton, and Prada were buying and selling products that were being produced on deforested lands. They also accused the government of turning a blind eye toward Amazonia, ignoring everything from deforestation to rampant land speculation, to granting amnesty to land thieves to allowing ranchers to employ slave labor. Around the same time, the International Finance Corporation (IFC) of the World Bank cancelled a multimillion-dollar loan to a Brazilian cattle company that had planned on using the money to expand cattle ranching in the Amazon.

With so much at stake, the government and the private sector met with the nongovernmental organization (NGO) community to hammer out some guidelines. One result was the "Beef Moratorium Protocol" in which signatories agreed not to purchase beef from recently deforested areas in Amazonia. This progress toward a more sustainable path served as an important precedent, and the Brazilian government augmented monitoring and protection of the Amazon forest. At the same time, efforts have been made to improve monitoring and certification of cattle and ranches, although the economic decline in Brazil has jeopardized much of the progress that was made.

The Brazilian NGO IMAZON, best known for its cartographic work, has been spearheading an innovative approach to managing the growing challenge in a constructive manner.

While there are more than 400,000 ranchers in the Brazilian Amazon, IMAZON's investigations revealed that just 128 slaughterhouses process 93% of the cattle. They propose that focusing on effective deforestation enforcement partnerships with slaughterhouses will prove more effective—and more cost effective—than trying to monitor and enforce extant legislation in hundreds of thousands of operations.

The stakes could not be higher. Brazilian sociologist Luiz Barbosa recently sounded a less optimistic note: "There will be no future for Brazilian Amazonia unless cattle herding is kept under control in the region."[1] With a national government actively promoting the spread of cattle ranching further into the rainforest, more than 74,000 fires were documented in the Brazilian Amazon in 2019.

At present, cattle ranching represents the leading cause of deforestation in other Amazonian countries as well, although I focus here on Brazil because it has the largest cattle herd and because it has the best data available. As the other countries are connected by better roads, the industry will most likely increase sharply. Of the nine Amazonian countries, only French Guiana and Suriname have chosen to forgo the development of a rainforest-based cattle industry.

What is the status of hydroelectric dams in Amazonia?

There are few communities on earth not thirsting for additional electrical power, from corporations running massive cloud server arrays to indigenous peoples in the remote Amazon hoping to charge a few mobile phones. In northern South America and most of the rest of the tropical world, population growth and urbanization drive this demand. Whether the direct objectives are poverty alleviation, energy security, or exports of energy-intensive commodities such as aluminum, generating more and cheaper electricity is thus a priority for the governments of these countries, all the better if doing so contributes to climate change mitigation.

For more than half a century in Amazonia, beginning with the damming of the Suriname River to create the Afobaka Dam in 1963, South American governments have largely deemed hydroelectric dams a seemingly limitless source of clean, renewable electrical energy, achieved without the burning of fossil fuels. More recently, however, as the negative impacts of these dams have become more apparent, many have come to question the desirability of both existing dams and the myriad of proposed projects.

There are many large hydroelectric dams operating in Amazonia, the majority in Brazil, with many more in various stages of planning. For a variety of reasons—mostly poor planning, corruption, overestimates of output, and underestimates of costs—these dams often operate below capacity and seldom generate the predicted return. A 2014 analysis by the Saïd Business School at Oxford found that the Itaipu Dam in southern Brazil cost 240% over the original estimate, sizeable enough to generate a negative effect on the country's public finances for three decades. Though it generated enormous amounts of electricity, "it will likely never pay back the costs incurred to build it."[2] The Oxford study—which analyzed 245 dam projects—concluded that construction costs on large dams are more than 90% over their original cost estimates and that "large dams in a vast majority of cases are not economically feasible and . . . risk drowning their fragile economies in debt."

As we have seen, fish are a major protein source in Amazonia, so protecting and augmenting their well-being should be a priority goal for all government planners, but it is precisely these creatures that inadequately planned and constructed hydroelectric dams most directly threaten. Dams have disrupted the migration of some commercially important species, particularly the giant catfish (*Brachyplatystoma filamentosum*) in the Madeira River. Dams are also known to negatively affect the annual water cycle in floodplain forests where many important commercial species spawn. They also alter the downstream movement of Andean sediments, which in turn can negatively impact fisheries as far east as the Amazon estuary and even out into the Atlantic. Moreover, the changes in water temperature brought about by dam power generation may diminish fish populations, not least the vulnerable hatchlings. Furthermore, the dams also obstruct the movement of manatees and dolphins, undoubtedly affecting this complex ecosystem in ways still incompletely understood.

Large dam projects also feature in the unsavory history of the displacement of Amazonian forest peoples, such as Suriname's Saramaka Maroons, exiled from the heart of their traditional lands. Today, traditional groups such as the 14 Amerindian tribes of Brazil's Xingu, who face disruption of their cultures and lands by the Belo Monte and associated planned upstream dams, pay the largest human price for their construction. Estimates are that Belo Monte may end up displacing 20,000 people.

It is not just the diversion and alteration of rivers that hurt indigenous and other rural populations: water quality declines while the roads built to cart in construction materials encourage settlement by outsiders and drive deforestation. Itinerant workers employed in construction often introduce disease and pollution, bringing further misery to indigenous communities and other peoples that remain in the vicinity. Dam construction necessitates the creation of quarries and borrow pits that can serve as breeding pools for malaria-carrying mosquitoes.

While such mega-projects are often sold to the public as emblems of national pride and vital tools for energy security, the ultimate beneficiaries are not the marginalized members of national society but the wealthy and powerful mining and mineral export industries. The Brazilians came up with a brilliant descriptive phrase for these types of costly mega-projects initiated by and primarily for the ultra-elite: *obras faraonicas*—"pharaonic works."

Analyzing the costs and benefits of Belo Monte, Dr. Philip Fearnside, the leading expert on Brazilian dams, noted,[3]

> The social benefits obtained in exchange for the dam's impacts are much less than official statements imply because much of the energy would go to subsidizing the profits of multinational aluminum companies that employ a miniscule workforce in Brazil. For example, the Albras smelter at Barcarena, Para, employs only 1200 people, but it uses more electricity than the city of Belém with a population of 1.2 million.

Compounding the damage, the artificial lakes established behind the dams flood and destroy enormous swaths of rainforest. Once considered sources of green energy, these dams are increasingly recognized as greenhouse gas generators. Not only is carbon released from the soil and dying vegetation once a forest is flooded, decaying vegetation in a low-oxygen environment at the bottom of the reservoir produces titanic quantities of methane—so much so experts term these dams "methane factories." One recent estimate is that these reservoirs release as much methane into the atmosphere as the entire aviation industry.

A primary reason these reservoirs are so large is that they are so flat—in other words, dams built on more of a slope can generate more energy in a smaller area—and Brazil represents (for the most part) the flattest part of the Amazon Basin. Smaller and better-placed dams could have less negative impact, especially if planners consider the complex ecological interconnections that characterize Amazonia. What is critical is strategic planning that examines and evaluates impact based on a single river system or even a single country. Better and more accurate longer-term cost/benefit analyses are also essential, especially in light of a changing climate and diminished forest that make warmer and drier conditions highly likely.

Finally, Amazonian countries do need to evaluate the possibilities of attaining energy security without fossil fuels or the construction of new mega-dams. The nine Amazonian countries are located on or near the equator. Meanwhile, the cost of solar and wind-generated electricity will continue to plummet. As such, the Amazonian countries are perfectly placed to develop renewable energy sources both for their own people and for sale to other countries.

What is the impact of gold mining in Amazonia?

The Incas had a massive appetite for gold, employing it for personal adornment, creating exquisite artworks, and utilizing it to decorate their temples. Much of their gold was extracted from the Andes, but at least some was mined in the Amazon: Incan roads penetrate as far as the Colorado River in the southern Peruvian Amazon, still a major focus of mining operations.

In one of premodern history's great ironies, however, none of the Amazon's earliest European explorers who came in search of gold encountered any sizeable quantities of the precious metal, despite the existence of what we now know to be substantial gold deposits. With little more than shovels and pans, Europeans began mining gold in the Amazon as early as the 16th century. Today, massive amounts are being extracted from Amazonia: Peru, for example, is the world's sixth largest gold-producing country. Even in little Suriname—the smallest country in Amazonia, about the size of the state of Georgia—gold purchased from small-scale miners by licensed buyers is worth nearly a billion dollars annually, and this does not count unregulated transactions.

In the early 1990s, the Brazilian government took note of the increasingly destructive impacts of gold mining in the rainforest and began to crack down on this damaging practice. Many of the miners—known as *garimpeiros*—fled across the borders into neighboring countries where environmental controls ranged from slim to nonexistent. The spike in gold prices after the 2008 global financial meltdown also created a gold rush, luring desperate people into the rainforest in search of riches.

Serra Pelada is an enormous gold mine approximately 270 miles (434 km) from the Amazon River mouth. A Brazilian farmer stumbled across gold on his property in 1979, word spread rapidly, and the rush was on. At its peak in the early 1980s, as many as 100,000 *garimpeiros*, looking like so many ants as they dug, struggled and fought in the open pit. The nearby shantytown was said to suffer 80 unsolved murders per month. Declining yields, pollution, and flooding led to abandonment of Serra Pelada in the late 1980s.

Unfortunately, gold mining has destructive impacts on the forest, on rivers, on animals (particularly fish), and on the local populace (particularly the indigenous peoples). Miners use high-powered hoses to disintegrate riverbanks and heavy machinery to excavate gold-yielding gravels. Compared to cattle ranching and large-scale agricultural operations, mining is not a major cause of deforestation in most of the Amazon, but it represents the major driver of forest destruction in the nation of Guyana as well as destruction of riverbank forests throughout the Amazon Basin.

Two highly toxic chemicals play vital roles in the process of gold extraction. Large-scale operations often employ cyanide, and these large companies tend to be relative better stewards of this poisonous material. Nonetheless, in a notorious 1986 incident, more than 400 million gallons (1,514 million L) of cyanide-laced mining wastewater was accidentally released into the Omai River in Guyana, with serious and long-term negative impacts on water quality, fish stocks, and both Amerindian and urban populations who suffered serious nerve damage and accelerating rates of cancer.

A more pernicious phenomenon is the widespread use of mercury in smaller-scale mining. Mercury, a powerful and persistent neurotoxin, is widely employed to amalgamate gold. It is estimated that for every gram of gold collected through small-scale mining, more than a gram of mercury is released into the environment. Some is burned off, and it then enters the atmosphere and can return in precipitation. The rest of the mercury is released into waterways, where it settles into river sediments and ends up in the food chain.

Accumulating in the human body, mercury poisoning causes irreversible nerve damage, brain damage, retardation, and birth defects when absorbed by pregnant women. In 2012, investigations in Suriname yielded grim conclusions: more than 40% of the predatory fish (like the ayumara/traira, *Hoplias aimara*, a favored food species) in the nation's rivers contained higher mercury levels than the European Union considered suitable for human consumption. In the southeastern corner of the country, hair samples of almost 80% of the children in forest communities contained elevated mercury levels.

Indigenous peoples in Amazonia tend to be heavily victimized by mercury pollution, particularly because fish serve as their major protein source. Due to their proficiency as hunters, Amerindians are often hired by gold-mining camps and serve as laborers as well. Needless to say, there are typically few practices in place to protect the indigenous peoples or other workers from mercury-based impacts. And the social impacts of these camps, which are often nests of pollution, prostitution, malaria, alcohol, illegal drugs, and violence, spread far beyond the confines of the camps themselves. In places as distant as Peru and Suriname, elevated mercury levels have been found in workers' blood and hair samples, as well as in samples from the residents of nearby villages.

A sad irony here is that low mercury and even mercury-free gold-mining techniques are known and available but have failed to become commonplace in the Amazon due to ignorance, greed, and the relatively low cost of mercury.

Gold mining may not even be in Amazon nations' ultimate bottom-line economic interests because—despite the tremendous social costs—most of the economic return is generated outside the region.

Gold has many uses around the world, from gold salts employed in the treatment of chronic arthritis to corrosion-resistant electrical connectors in high-tech devices. About half of all gold, however, is made into jewelry, and the biggest consumers of these ornaments are the Chinese and the (Asian) Indians. Destroying the rainforest of the Amazon and poisoning its aboriginal inhabitants seems like a considerable price to pay for personal adornment.

What is the role of large-scale agricultural production in deforestation?

In the Amazon, large-scale agricultural farms—usually producing a single crop (known as a *monoculture*) such as soy, oil palm, or sugarcane—have been responsible for the deforestation of vast areas, and this trend seems likely to accelerate. Ancient, weathered Amazonian soils have long been considered infertile, fragile, and incapable of supporting intensive agriculture. Recent agricultural research and experimentation based on the massive addition of fertilizers, vitamins, minerals, and other chemicals;[4] the introduction of new forage grasses; and new means to reduce pest management may be changing the paradigm. However, these novel approaches are highly capital-intensive, largely limiting their successful application to wealthy ranchers and landowners.

As a result, huge mechanized farms are becoming increasingly common and pushing small farmers deeper into the rainforest. Driving the economics of these large farms are low land prices, low labor costs, access to modern technology and capital, expanding road and utility infrastructure, fuel subsidies, low taxes and tax avoidance, state-supported agricultural research and extension, and the linking of these Amazonian farms and goods to the global market via development projects like the Initiative for the Integration of the Regional Infrastructure of South America (IIRSA), a program to better link all countries in terms of energy, telecommunications, and transportation. Furthermore, weak or nonexistent law enforcement favors wealthy and powerful landowners who may even chase smaller farmers off their land.

Soy has become the major agricultural export commodity of both Bolivia and Brazil as demand continues to grow because of recent confused US tariff policies. Though soy is most easily cultivated in temperate or subtropical regions, the development of new varieties adapted to the tropics in the 1990s, coupled with the application of massive quantities of agrochemicals, enabled soy cultivation to expand rapidly in Brazil; greatly increased deforestation and water pollution ensued. Soy agriculture is capital-intensive but not labor-intensive, meaning that only wealthy ranchers can afford to run these highly mechanized farms, and they employ relatively few people.

A milestone in the evolution of the Brazilian soy industry was a scathing 2006 report by Greenpeace ("Eating Up the Amazon")[5] that linked both Cargill and McDonald's to deforestation for their role in expanding soy plantations in South America. The result was the Amazon Soy Moratorium (ASM), in which major soy traders pledged not to purchase local crops produced on recently deforested lands, and which banned direct conversion of Brazilian Amazon forests to soy fields. It represented the very first zero deforestation agreement in the tropical world. The Brazilian government played a positive role in this process by enacting measures to monitor compliance both on the ground and via aerial imagery, and by going so far as to deny credit to areas that failed to enforce the deforestation ban. More than a decade later, estimations of its achievement are contested: while corporate participants like Cargill have proclaimed the ASM "a resounding success," others have concluded that the documented reduction in deforestation rates was primarily due to other factors. And some of the corporate giants have either missed their targets or begun backing away from them in light of the current Brazilian government's pro-development policies.

The other foreign plant expected to supplant vast stretches of the Amazon rainforest is oil palm. Native to western and southwestern Africa, this palm yields substantial quantities of two oils, one from the fruit and one from the kernel. Prized for their shelf stability, these oils can be processed and used separately or together in a staggering variety of products—indeed, they are the most widely used vegetable oils in the world and may be found in as many as half of all consumables on conventional supermarket shelves, including baked goods, chocolate, ice cream, margarine, cosmetics, detergent, pharmaceuticals, shampoo, and soap. They are also used as biofuels. Concomitantly, global demand for oil palm is expected to increase by 50% in the next 15 years.

Oil palm first came to Brazil as part of the slave trade in the 16th century. Locally known as *dende*, the oil has always served as an essential component in the cuisine of the state of Bahia, the most Afro-centric state in Brazil. The initial commercial plantings in Amazonia were near Belém in 1974, but it was not until a quarter-century later that the Brazilian government began promoting it as an appropriate large-scale crop for Amazonia. Oil palm is now widely planted in Pará; overall, oil palm acreage in Brazil doubled between 2004 and 2010 and is projected to double again in the next decade.

Elsewhere, more than 4,000 square miles (10,360 km^2) are already being farmed in Colombia and Ecuador, and plantations are also expanding in the Peruvian Amazon. Currently, the world's major producers are Indonesia and Malaysia, where the cultivation of oil palm monocultures has been the direct cause of extensive deforestation. Correspondingly, oil palm represents the commercial crop most likely to expand in Amazonia.

Because of its high productivity, and because it requires more labor than soy plantations, some have suggested that oil palm would be an excellent crop for small farmers, particularly if it could be part of an effort to convert degraded cow pastures into productive lands. Such an initiative would be worthwhile if carried out successfully with the correct technical expertise and sufficient funding. Many experts worry, however, that "greenlighting" oil palm expansion in Amazonia would more likely result in massive clearing of rainforest.

What is the role of small-scale farming in deforestation?

"Small-scale agriculture" refers to growing crops primarily for a family's own consumption or for sale in local or national markets, rather than primarily for national or international markets. In the Amazon, these farms tend to be smaller—often much smaller—than a square mile (2.6 km^2).

Compared to large cattle ranches and mechanized monoculture plantations, such farms are not in and of themselves direct drivers of major deforestation in the Amazon. However, they can play a major role as the point of the spear—or the edge of the machete—in initiating forest felling in previously pristine areas. When soil quality declines and crop yields wane, a small farmer may abandon his original plot and fell additional forest. The vacated lands are often absorbed into cattle ranches or larger farms whose owners possess the capital and/or expertise to extract profit from degraded lands.

As mentioned earlier, slash-and-burn agriculture is how indigenous populations in Amazonia "farm the forest." When a relatively small population utilizes this approach in an enormous rainforest, this process can continue seemingly indefinitely. As the population of the Amazon continues to increase, however, this traditional system becomes less common and sustainable.

Who farms in Amazonia?

The Amazon is home to many peasant peoples who are descendants of tribes that no longer exist. Groups like the *caboclos* of the Brazilian Amazon have sometimes been called "neo-Indians" because so many of their cultural and agricultural practices echo those of their indigenous forebears. One measure of the extraordinary agricultural sophistication of indigenous cultures is the diversity of crop varieties. For example, anthropologist Janet Chernela recorded 135 varieties of cassava in the gardens of one tribe in the Colombian Amazon. By comparison, agricultural plots of recent immigrants to Amazonia are more likely to contain fewer than a dozen.

At the other end of the spectrum with respect to agricultural sophistication are recent desperate immigrants to the Amazon rainforest from other regions, such as peasants abandoning drought-ravaged regions in northern Brazil, Afro-Americans fleeing civil unrest in coastal Colombia, or highland indigenous peoples attempting to escape grinding poverty in Bolivia or Peru. Having little knowledge of local agricultural practices, training, government support, or access to credit, their likelihood of success is limited, and they are unlikely to practice sustainable cultivation. Furthermore, as they largely lack legal title to their lands, they may be coerced by neighboring large landholders into abandoning their lands, thus driving them deeper into the rainforest. These land barons may pay them a pittance to sell or—in some cases—have hired thugs to simply chase off the small farmers and then seize their lands.

One hopeful initiative focuses on land tenure reform. Lacking legal title to their lands, small stakeholders tend to deforest their land. However, when they hold legal title to their lands and can count on the government protecting them from large landholders attempting to chase them off, these people become better stewards of their natural capital, and deforestation rates decline.

What is the status and impact of logging in Amazonia?

Loggers have long looked at the Amazon rainforest in the same way that the first Western settlers on the US Great Plains viewed the buffalo: as an inexhaustible resource to be plundered with impunity. The results in some corners of Amazonia—like parts of the Colombian Putumayo in the northwest or the Brazilian state of Para in the northeast—have been the same: the seemingly limitless resource has been extirpated.

Compared to large-scale agriculture, cattle ranching, and mining, logging is a lesser cause of deforestation. Most logging in the Amazon has been along river courses or roads. For a variety of reasons, Amazonian wood has relatively low commercial value; even a mature tree of the most valuable species—mahogany—sells for only a few thousand dollars. The majority of logs harvested from the Amazonian rainforest are used locally or domestically (for example, felled in the Amazon and then shipped to Rio de Janeiro or São Paulo as building material); relatively few species are exported and then often as raw logs, further diminishing their value to the country of origin.

Despite the development of increasingly sophisticated forestry techniques worldwide as well as increasing global concern over the fate of tropical forests, many practices and trends in Amazonia remain harmful to the ecosystem as a whole.

Fatefully, the great biodiversity of Amazonian forests presents special challenges in developing the means to harvest timber sustainably and economically. A typical forest stand will likely harbor only a few species of relatively high economic value. Moreover, the characteristically poor soils of Amazonia are easily damaged, resulting in harm to the ecosystem as a whole from logging operations. In addition, the tendency of rainforest lianas to effectively bind multiple trees together means felling a single tree may result in collateral damage to numerous adjacent trees.

The standard approach to timber harvesting in Amazonia has been selective logging: identifying and felling the few individuals of the most economically valuable species. Recent studies, however, reveal that this method has proved much more damaging to the residual forest than was previously believed. Ecologist Greg Asner concluded that such operations lead to multiplied gaps in the forest, increased forest fragmentation, more light reaching the forest floor, increased likelihood and severity of fire, and perturbations in nutrient cycling and hydrological cycles and other fundamental ecological processes.

The most pernicious legacy of selective logging is that it tends to open the forest up to further human disturbances: incursions and deforestation. Increasing access to the forest by building roads and logging camps, or opening the forest by removing large trees, often takes a subsequent heavy toll on the ecosystem: legal operations at the outset can open the door to illegal and massively destructive operations. Research has shown that selectively logged forests are more vulnerable to fires. One study concluded that opening the forest to selective logging results in complete land use change—which can be total deforestation—in one-quarter of the cases.[6]

The system is both plagued and characterized by massive illegality. There exists a vast mafia involved in falsifying extraction quotas of legal operations to "greenwash" illegal harvests. An analysis in the Brazilian state of Para found that 78% of the logs checked were harvested illegally. The best-monitored forestry in Amazonia is in French Guiana, which features the smallest land area and population in tropical South America and is thus (theoretically) the easiest to control. In sharp contrast, illegal logging is rampant in parts of Bolivia, Brazil, Colombia, Ecuador, and Peru.[7]

Once logging takes hold, it often metastasizes out of control, especially in countries with chronically understaffed departments overseeing forestry operations and environmental protection. The town of Sinop in the Brazilian state of Mato Grosso was founded less than 60 years ago. Even though it has fewer than 150,000 inhabitants, forester Dominiek Plouvier (then of the World Wide Fund for Nature) found more than 300 sawmills operating in the 1990s. The name "Mato Grosso" means "Great Forest" but, by the early 2000s, *no* large stands of rainforest remained in the state outside the Xingu Indigenous Reserve since the forest had been cleared both for agriculture and to feed the sawmills.

As the global population continues to increase, global demand for timber is expected to rise. According to a 2016 report by Interpol and the UN Environment Programme (UNEP), illegal logging around the world already represents the world's highest-value environmental crime—estimates are as high as $152 billion per annum—and many major crime syndicates are deeply involved in this loathsome trade. In addition to these non-state actors, China is a major conduit for illegal timber, primarily from Southeast Asia but with a global reach throughout the tropics.

Even the best-intentioned efforts can produce unforeseen circumstances. Ethnobotanist Glenn Shepard has noted that when the Brazilian government banned mahogany exports in the mid-2000s, there was a massive mahogany boom in adjacent Peru. This led to loggers invading the Piedras region, which displaced many members of the Mashco Piro tribe, who sometimes responded aggressively. Mashco Piros are sometimes now encountered living inside legal logging concessions, demonstrating that the cascading effects of extractive industries go beyond what may be visible at their worksites.

One attempt to manage or reduce this illegal trade was the founding of the Forest Stewardship Council (FSC) in 1993 to certify sustainably produced timber, thereby augmenting the economic incentive to manage forests (particularly tropical forests) sustainably. A decade later, the European Union created the Forest, Law, Government and Trade Action Plan (FLEGT) to similarly battle the global scourge of illegal logging by improving sustainable, efficient, and legal forest governance and management. FLEGT noted that illegal trade was not only destroying tropical forests, diminishing biodiversity, and contributing to climate change but also costing hard-pressed tropical governments billions in revenues. In March of 2018, Greenpeace announced that it would not renew membership in the FSC, claiming that the FSC was not meeting its aim of protecting forests and ensuring that human rights are respected.

How will climate change affect Amazonia and vice versa?

Human activities pump greenhouse gases like carbon dioxide into the atmosphere, driving climate change. After fossil fuel consumption, deforestation—primarily in the tropics—is the second largest source of these emissions. Per recent estimates, approximately 10% of all global greenhouse emissions stem from forest destruction and land use change.

Alterations predicted by modeling of the Amazonian climate include rising temperatures, diminished rainfall, more frequent droughts, and variations in seasonality, such as the early arrival of, delay in, or elimination of the rainy season. These changes will transpire in tropical ecosystems populated by many plants and animals poorly adapted to manage these perturbations. Unlike temperate organisms, tropical species are not accustomed to major seasonal variations in temperature. Over the past few decades, many species have moved toward the poles and/or higher in elevation in search of cooler climes. According to some experts, however, average temperatures are roughly equivalent within 10 degrees of the equator, and thus much of the Amazonian flora and fauna will be unable to migrate to an optimal thermal comfort zone. In the far west of the Basin, cooler average temperatures can be found by moving up the eastern slope of the Andes, but few rainforest species have the ability to propagate themselves in this way.

Moreover, many rainforest creatures like amphibians, insects, and reptiles are poikilotherms (cold-blooded animals), unable to control their body temperatures and therefore especially ill adapted to a changing climate. Even tropical species that possess some ability to regulate their body temperatures, like mammals and birds, have limited adaptability compared to their temperate counterparts: one tropical ecologist poignantly recorded his sadness while observing dying fruit bats in Australia plummeting to the ground from their arboreal perches during a brutal heat wave. Another biologist terms this lethal process—when a local climate warms beyond a species' ability to survive and no cooler clime exists within reach—"biological attrition."

Deleterious effects will impact Amazonian flora as well. In the 1970s, Brazilian climatologist Eneas Salati began to decipher the complex interrelationship between Amazonian vegetation and climate. Contradicting the previous perception that the rainforest is merely a passive recipient of precipitation, Salati's experiments demonstrated that the Amazon creates half of the rainfall it receives by recycling the water as many as six times as air and cloud masses move west from the Atlantic to the Andes.

Trees suck water from the soil and release it into the atmosphere through transpiration—a single tree can emit as much as 80 gallons (302 L) in a 24-hour period. Canopy leaves emit volatile substances that serve as nuclei for water vapor and precursors of raindrops. The *biotic pump* theory—coined by A. Makarieva and V. Gorshkov—hypothesizes that the huge amounts of water vapor emitted into the atmosphere above the rainforest reduce atmospheric pressure, which in turn draws in additional moisture from the ocean, producing more precipitation over the forest. Furthermore, Brazilian climatologist Carlos Nobre notes that excess water vapor is transported as "aerial rivers" to produce abundant rainfall in other regions as well.[8] Fewer trees means less rain in Amazonia and in adjacent regions like São Paulo state (Brazil's industrial heartland), southeastern Brazil (Brazil's agricultural breadbasket), and parts of Paraguay, Uruguay, and even Argentina.

Meanwhile, the constant influx of freshwater into Amazonia also provides vital services for local people. Fish remain the major protein source for most rural inhabitants and many urban dwellers, as well as a major source of income. Amazonians use their streams and rivers as sources of potable water and for fishing, bathing, and travel. Bountiful rain extinguishes forest fires, generates power by turning the turbines of hydroelectric dams, and makes both subsistence and industrial agriculture possible.

I was in Acre state in western Brazil during the crippling drought of 2005. Millions of trees died throughout the Basin. Major rivers were reduced to stinking puddles of dying fish. Hunger was rampant. River transport came to an absolute halt. Fires burned out of control. Thick plumes of smoke filled the air, dehydrating the atmosphere and creating giant rain-shadows downwind of major forest fires that burned out of control. The ubiquitous smoke reduced visibility to zero and airports closed while hospitals filled with patients suffering and even dying from smoke inhalation. The economic costs and the toll of human suffering were incalculable. Regionally, going forward, the crisis of 2005 likely will not be singular: 2015 was warmer still.

Climate models predict a grim future for the Amazon. Warming ambient temperatures, warming water temperatures, decreased precipitation and less predictable seasonality, and more extreme weather such as heat waves, droughts, and floods seem likely. Also problematic are feedback loops in which interlocking factors magnify negative effects: increased forest fires eliminate canopy cover, allowing additional sunlight to desiccate the forest floor vegetation, thus enhancing its flammability. Forest fragmentation exposes more of the ecosystem to the sun's drying rays around the now-exposed peripheries. Large tracts of once-moist forest will parch due to reduced transpiration, which further diminishes rainfall and intensifies drought, thereby making trees more flammable—a phenomenon known as the "vegetation breeze." Road building, logging, land clearance for industrial agriculture, and the replacement of wet forests by drier grasslands augment the likelihood of further forest fires, carbon release, and destructive climate change. And the human inhabitants will also suffer. Climate change will increase mortality due to longer and stronger heat waves, declining air quality, additional drought, and reduced water quality and food security.

Amazon experts warn of a deforestation tipping point, a situation beyond which so much forest cover has been removed and so little rain is generated that a majority of the once-great Amazon ecosystem is reduced to a dry and depauperate grassland.

What is the Trans-Amazon Highway, and how has it driven deforestation?

Deforestation is a complex issue with many causes, some of which we have covered. Overpopulation, social inequality, lack of economic opportunities, and local, national, and international demand for commodities all play a role in the ongoing destruction of the Amazon. Cattle ranching, small- and large-scale (particularly soy and oil palm) agriculture, logging, mining (particularly for gold), road building, narcotrafficking, petroleum exploration and extraction, and poorly planned colonization schemes all are drivers of the deforestation process. Unclear tenure regimes, weak governance, widespread corruption, limited monitoring and enforcement of forest protection, and deforestation-friendly agricultural subsidies and/or tax policies encourage destructive policies and practices.

The attendant destruction of biodiversity disrupts the complex and interdependent web of life that is the rainforest. Local human communities, part of that web, face multiple challenges, particularly in the form of reductions in edible fish species and water quality, the latter caused by increased erosion and toxic runoff from mining operations. Additionally, there is a growing consensus that mass deforestation in Amazonia affects both local and regional climate—in 2005 and 2010, the Amazon region suffered two extremely severe droughts, events that even impeded planes from arriving and departing due to heavy smoke cover from burning forests. The economic costs of deforestation, we are finding, may be both unpredictable and severe.

In a 2015 study, the World Wildlife Fund identified 31 "deforestation fronts,"[9] where forests were being felled or degraded at a rapid pace. The most threatened zone was the so-called

Arc of Destruction—the southeast Amazon. The other recognized deforestation fronts are located primarily around the edge of the Amazon, although a few were designated in the central Amazon around the city of Manaus or along the lower stretches of the Amazon River.

Until the 1970s, there was little ecosystem destruction in Amazonia. Then the Brazilian government commenced implementation of plans to "develop" the Amazon. Of course, the predominant original European settlers were the Portuguese, a seafaring race who had colonized and settled mostly on or near the Atlantic coast. In 1960, the nation built an inland capital, Brasilia, to encourage and facilitate the peopling of the interior.

The Brazilians also launched an ambitious program of road building into the interior—the Trans-Amazonian Highway—which moved urban dwellers into the Amazon as never before. This resulted in a number of unforeseen problems: mass migration into the forest by farmers unskilled in rainforest agriculture, land speculation, illegal logging, violence, genocide of tribal peoples, and wildcat gold mining. Several decades of wanton forest destruction ensued.

As the many negative effects of this poorly planned initiative became increasingly apparent, a local and international cry was raised by those denouncing environmental destruction for the economic benefit of the very few. Numerous environmental and indigenous NGOs formed within the Amazon countries, visionary and charismatic local leaders—both rural, like Chico Mendes, and indigenous, like Paiakan and Raoni of the Brazilian Kayapo tribe—joined forces with international celebrities like Sting to issue stirring calls for better environmental stewardship, respect for human rights, creation of protected areas, and recognition of indigenous land rights. The late 1990s and early 2000s saw improved conservation laws and law enforcement, enhanced monitoring through technology, international campaigns, ongoing pressure from civil society, and the continued rise and power of local and international conservation organizations.

One of the great success stories in the relatively brief history of Amazon conservation was a 75% reduction in deforestation rates in Brazil—the nation with by far the most Amazon forest—between the years 2000 and 2013. Soon thereafter, economic hard times and insufficient on-the-ground enforcement of existing legislation led to an 80% spike in the annual deforestation rate. According to Phil Fearnside, a leading authority on the Brazilian Amazon, the passage of a new Forest Code in 2012 weakened critical environmental protections and offered an amnesty to those who had violated environmental laws prior to 2008. Furthermore, the August 2016 impeachment of Brazilian president Dilma Rousseff was believed to have been spearheaded by the extractivist and agribusiness industries so they could fell more forest. As of this writing, the deforestation rate remains troublingly high and appears to be increasing.

Nor is Brazil the only source of concern. Venezuela was once widely regarded as having one of the best-designed and best-protected national park systems in South America. However, rampant corruption and mismanagement have brought the country to its knees. Not only are people fleeing into the forest in search of food and gold, but the country's political leaders are attempting to grant mining and forestry concessions at bargain-basement prices—with little or no social or environmental controls.

Colombia long featured the lowest deforestation rates in Amazonia. Ironically, the country's terrible civil war featured an environmental upside in terms of keeping mega-development projects, corporate enterprises, and road building at bay. Since the signing of the historic peace accords in 2016, deforestation has spiked, increasing nearly 20% during the following 2 years. Unlike in Brazil and Venezuela, however, the federal and state governments have proved to be outspoken proponents of good environmental stewardship—government officials have pledged to achieve zero net deforestation in the

Colombian Amazon by 2020. Time will tell if they will be able to convert those positive sentiments into positive actions in and for the rainforest.

What is the impact of oil and gas exploration and extraction on local peoples?

That South America is a continent rich in petroleum was noted early. The first Spanish explorers who arrived at Lake Maracaibo on the Venezuelan coast in 1499 found the tribal peoples there employing the oil that seeped from the ground surrounding the lake as medicine to coat their wounds (a precursor of Vaseline) and as pitch to coat their torches (a precursor of the petroleum industry). The first commercial extraction of oil from the Amazon did not happen until almost half a millennium later, when Texaco began pumping oil from the northern Ecuadorian Amazon in the mid-1960s.

The Andean Amazon region is rich in oil and gas, but access and production have long been hindered by the region's relative isolation, high rainfall, challenging topography, suspicious tribes, and—in the case of Colombia—civil war. Until the recent cessation of hostilities, Colombian guerrillas would regularly sabotage pipelines to harass the government or threaten to destroy pipelines in order to extort money.

When the oil industry moved into the northwest Amazon, environmental legislation (by the national government) and environmental concerns (by the international community) were all but nonexistent except for the indigenous and peasant communities living nearby or downstream. The result was hideous pollution: it is alleged by some environmental groups that Texaco poured 17 million gallons of crude and as much as 20 billion gallons of toxic byproducts into local rivers in the Ecuadorean Amazon. And there have been major oil spills in Bolivia, Colombia, and Peru as well.

The toxic legacy created by such negligence has proved to be widespread and deadly: contaminated lagoons, lakes, rivers, and streams. Drinking water has been poisoned, and soils and gardens have been fouled. Even game animals like tapirs have been observed consuming toxic waste. The epidemiological impact on human communities is predictable: gastric ailments, miscarriages, skin diseases, cancer, and death.

The IIRSA initiative mentioned earlier is a regional development effort, one focus of which is building new roads deeper into the rainforest with the intention of linking the Amazon to ports on both the Atlantic and Pacific coasts. And growing concerns over climate change have given new impetus to the search for natural gas, which burns cleaner than oil and is abundant in Amazonia. At the same time, the instability of oil markets and prices—a barrel of oil has risen and fallen in value from $24 in 1998 to $150 in 2008 to $50 at the time of this writing—created a disincentive to invest heavily in the infrastructure needed to extract and export petroleum products. Global concerns over climate change provide further disincentives.

The Camisea Natural Gas Project is emblematic of the many problems that can and do result from these projects in Amazonia. This project—encompassing the largest natural gas reserves in the Americas—was launched in 2000, in the Urubamba Valley near Machu Picchu in the Peruvian Amazon. The effort has yielded billions of dollars, but, particularly during the early years, it is believed to have introduced infectious diseases like respiratory ailments that attacked vulnerable members of isolated tribes. Like almost all oil and gas projects, Camisea built roads into the forest that provided access to peasant settlers and land speculators who felled forest. The inevitable soil and water pollution, erosion, landslides, and diminished fish populations have led to hunger and malnutrition. Though widely touted at the outset as a positive example of controlling deforestation by limiting road building, the pernicious effects on game populations (including fish) and indigenous well-being are increasingly obvious.

Many complain that the project has generated a host of negative downstream effects, from increasing child malnutrition to rampant corruption to deaths of isolated peoples from introduced infectious diseases.

An evaluation by the World Wildlife Fund concluded that the Camisea Natural Gas Project revealed several challenging lessons (which presumably hold true for many if not most projects of this kind in Amazonia), first and foremost that the national and regional governments lack the expertise and on-the-ground presence to ensure that companies prevent or minimize environmental and cultural impacts. This is further exacerbated by laws (as we have seen elsewhere in Amazonia) riddled with loopholes. Furthermore, the government often possesses neither the ability, the tools, nor the inclination to compel companies to obey the law. And neither the tribal, nor the local, nor the national governments possess the training and abilities to effectively plan for, manage, and spend the monies generated by projects of this type.

Several of these South American projects have yielded massive sums; some of these funds have generated positive effects in terms of funding for hospitals and schools. Typically, however, relatively little largesse makes its way to the communities most affected by the oil industry—the major benefits often accrue to the wealthiest citizens with the most powerful political connections. And, regardless of whether tribal or peasant communities are able to manage the funds efficiently and effectively, corruption and inefficiency divert and consume enormous amounts of the tax or concession monies paid by the oil and gas companies.

In response to widespread protests, several large companies have made concerted efforts to minimize negative impacts, such as by conducting research from helicopters and small planes, excavating fewer exploratory wells, and constructing fewer roads. However, so-called *second-tier companies*—smaller and regional—are moving into the industry along with foreign companies from countries with less vocal environmental constituencies (like China). Unless local constituencies in the Amazonian countries—people in both forest and urban communities—can make themselves heard in terms of enacted new legislation and monitoring and enforcement, we are facing more deforestation, more pollution, and more unnecessary human suffering.

Camisea seems a case in point. While the project has its defenders for making repeated efforts to learn from previous mistakes made by other mega-development rainforest projects, it has also attracted opprobrium. A recent visit by an evaluation team found repeated cases of oil spills, fish stock declines, inoperative installed water systems, and payments made to local communities without adequate training, planning, and guidance.

What is the impact of overhunting and overfishing in Amazonia?

On April 6, 1801, while ascending the Orinoco in central Venezuela, Alexander von Humboldt and Aimé Bonpland paused their journey to witness an annual harvest of turtle eggs. Just a month earlier, on three islands slightly north of the Atures rapids, hundreds of thousands of side-necked *urrau* turtles (*Podocnemis expansa*) had emerged from the river to deposit more than 100 eggs apiece. Hundreds of indigenous peoples from different tribes—each group distinguishable by their particular red and blue body paint—had encamped there to excavate and harvest the eggs. Some would be consumed on the spot but most of the eggs would be rendered into oil, which—at the behest of the missionaries—would be sold or traded for cooking or for lamps. Von Humboldt was told that Jesuit missionaries had previously overseen the commercial harvest and had insisted that a portion of the eggs be left undisturbed to hatch. However, the Franciscans had taken over and had neglected to maintain any such stewardship. Von Humboldt estimated that more than 30 million eggs were being harvested from the lower Orinoco. As a result—more than two centuries ago—the indigenous peoples reported that the egg yield was declining.

A half-century after von Humboldt's sojourn on the Orinoco, an equal if not more egregious abuse of wildlife was reported from the Brazilian Amazon by Henry Walter Bates:[10]

> It is scarcely exaggerating to say that the waters of the Solimões [upper Amazon] are as well stocked with large [black caiman] in the dry season as a ditch in England is in summer with tadpoles. . . . During a journey of five days I once made in the Upper Amazons steamer, in November, [black caiman] were seen along the coast almost every step of the way, and the passengers amused themselves, from morning till night, by firing at them with rifle and ball. They were very numerous in the still bays, where the huddled crowds jostled together, to the great rattling of their coats of mail, as the steamer passed.

Today, there exists no great ingathering of nesting turtles on the river islands of the central Orinoco, nor does one see any groups of black caiman on the banks of the Solimões. The net result of this senseless slaughter has implications for human well-being as well as for ecosystem form and function.

Black caiman were also subjected to an almost unimaginable slaughter. Though, as we have seen, the species is slightly smaller than the Orinoco crocodile to the north, their original range was many times greater, from the islands at the mouth of the Amazon like Marajo and Mexiana, west to the foot of the Andes. Enormous creatures, they were hunted both as a food source (primarily the tail) and as a source of edible fat. Because the animal was considered a major threat to cattle, caiman drives were held to destroy as many crocodilians as possible—one German biologist tallied 800 black caiman slaughtered over the course of 2 days. As Bates wrote, these great saurians in some instances were shot out of boredom or for target practice, not unlike the buffalo in the American West. But because of their immense size and extraordinary abundance, they served as species of choice for the skin trade. One biologist estimated that more than 5 million crocodilian skins were removed from the Brazilian Amazon each year in the early 1950s—and black caiman skins represented a sizeable percentage of that number.

Nor was the butchery restricted to Brazil. A Colombian biologist traveling in Guyana in the early 1970s recorded that rampant overhunting "resulted in such a wholesale slaughter that the lower reaches of the river stank for weeks . . . a single caiman hunter obtained 5,000 skins."

The zoological counterpart of deforestation is *defaunation*, the process whereby animal populations are reduced or eliminated. *Homo sapiens* has been hunting and fishing in the Amazon for 10,000 years, and, prior to the advent of Europeans, their weapons, and their capitalism, the majority of those efforts were sustainable. Most people lived in small and widely scattered, temporary settlements in the terra firme forest that characterizes most of Amazonia. Even when animals were overhunted in the vicinity of villages, there always existed vast tracts of forest and vast stretches of rivers, lakes, and swamps where animals could breed and populations would be replenished. And where human population density was highest—in the great indigenous villages along the main Amazon, as recorded by Orellana and Carvajal—we know that these peoples were raising side-necked turtles (*Podocnemis* spp.) in pens to help meet their protein needs. In fact, we now know that these creatures are relatively easy to raise as they feed themselves on aquatic vegetation and dead fish, reproduce prodigiously, and yield high-quality meat, eggs, and oil. But—unlike cattle—these turtles require no deforestation, antibiotics, or other expensive or toxic chemicals to survive and thrive.

When Francisco de Orellana descended the Amazon in 1541, the indigenous peoples hunted with blowguns, bows, arrows, and spears. Most animals killed in the forest had to be lugged back to the village where there were few means to preserve the meat, so hunters seldom dispatched more than could be consumed within the course of a few days. Tribal societies often possessed taboos against overhunting, which seemed to function as social controls designed to prevent overexploitation of both terrestrial and aquatic creatures.

All that has changed as these traditional forest societies have come into increasing contact with the outside world. Hunting calendars—which proscribed killing certain species at particular times of the year, often during their mating season—have gone by the wayside, frequently at the insistence of missionaries seeking to replace aboriginal beliefs with Western religion and the market economy. And not only are indigenous populations hunting and fishing unbound by any social controls: much of the ongoing population growth in Amazonia consists of migrants from coastal cities in Brazil or from the Andes who have virtually no spiritual attachment to the forest or the rivers that might mitigate overexploitation. Moreover, traditional hunting weapons have been replaced by the shotgun, which features greater killing power and requires substantially less skill.

Even more destructive than the introduction of guns was the advent of capitalism. Indigenous and peasant communities who until recently were isolated and harvesting resources for subsistence purposes are increasingly connected to regional, national, and international markets. By way of example, it was not very long ago that the acaí palm fruit from the Amazon estuary was a rare commodity in US retail markets.

With respect to market demand, however, it is defaunation that is causing the most concern, with special reference to the *bushmeat* trade, in which mostly large mammals (primarily tapirs, peccaries, deer, large primates such as spider monkeys and wooly monkeys, or sizeable rodents like capybaras and pacas) and birds (primarily guans and curassows) are slaughtered for sale in urban centers like Belém or Manaus, in mining camps, or even in tourism lodges. Better transport and refrigeration mean hunters and fishermen no longer must limit their take to prevent spoilage. The skin trade—cat pelts and caiman leather—as well as the pet trade have also taken a toll over the years. And, increasingly, folk medicine beliefs—especially in East Asia—are driving defaunation: recent reports from both Bolivia and Suriname are that Chinese buyers are financing the slaughter of jaguars for their teeth.

Such unrestrained practices have exacted a huge biological toll: the depletion of large animals, particularly birds and mammals, can disrupt a variety of interrelated ecological processes, leading to a series of cascading changes in animal and plant species composition. Such holds true for the fish as well: research initiated in the 1970s by ichthyologist Michael Goulding demonstrated that the well-being of the fish and the forest are closely intertwined. Sizeable fish species like the 60-pound (27 kg) tambaqui (*Colossoma macropomum*) facilitate tree reproduction by consuming, transporting, and excreting seeds. But because the tambaqui is one of the world's tastiest fish, overfishing has been so severe that it has impacted tambaqui populations more than 600 miles (965 km) from Manaus, a phenomenon that one biologist has termed a "defaunation shadow." The destruction of the tambaqui and other frugivorous fish has significantly decreased flooded forest seed disperser populations, a less obvious process that could add to deforestation.

If unabated overharvest of terrestrial and aquatic creatures continues, not only will the populations of those species decline, but other interlinked species will not thrive, will not be pollinated, will not reproduce, will not disseminate their seeds, and will cease to exist in that particular landscape, along with other animal species that depend on them.

In a classic 1992 paper, biologist Kent Redford memorably christened these zoologically depauperate ecosystems the "Empty Forest."[11] Redford wrote: "The presence of soaring, buttressed tropical trees, however, does not guarantee the presence of resident fauna. . . . The absence of these [large] animals has profound implications, one of which is that a forest can be destroyed by humans from within as well as from without."[12]

What impact is China having on Amazonia?

Within the course of several decades, China has risen to become the world's second largest economy. While much attention has focused on the impact of this growth on the industrialized world, this process is already producing environmental changes throughout the tropical countries of South America as China's aggressive, ambitious, and centrally organized leadership clearly sees Amazonia as a rich wellspring of resources ripe for harvest.

Since Orellana's initial voyage in 1541, the Amazon and its inhabitants have suffered impacts from the advent of many major foreign powers: Spain, Portugal, Holland, England, Canada, and the United States. But the scale of China's ambitions and power potentially outweigh the impacts of all who have come before, especially with the nation's ability to offer interest-free loans and its poor track record on both the environment and on human rights.

Not all is bleak on the Chinese environmental front. The government is making major strides to reduce or eliminate the hideous air pollution that has resulted from decades of rapid industrialization. Long serving as one of the world's biggest markets for elephant ivory, China announced a ban on all ivory and ivory products in 2017. And China is a proven leader in clean energy, not only in terms of the new technology it is pioneering, but also in being a signatory to the Paris Accords to combat climate change.

Meanwhile, the Belt and Road Initiative is a Chinese-led multibillion-dollar international development strategy. Per the Chinese, its purpose is to enhance regional and global connectivity; others perceive it as an attempt to achieve international dominance through a China-centered trading network. Though much of the effort has focused on Eurasia, the Chinese have also been aggressively expanding this program into sub-Saharan Africa and tropical America. In Amazonia, this has taken the form of infrastructure investment and construction (often with Chinese crews), including roads and hydroelectric dams, as well as mining and extraction of oil and gas.

China also has established the $20 billion Brazil-China Cooperation Fund for the Expansion of Production Capacity. Brazilian environmentalists decry this as an effort by both Brazilian and Chinese commercial interests to forge ahead with mega-development projects like enormous hydroelectric dams, railroads, and roads without adequate environmental or cultural protections.

In Ecuador, the previous president formed a close alliance with China in which China quickly became Ecuador's primary creditor and funder of an infrastructure boom in new dams, mines, and roads. Eight major dams were constructed, including the Coca Codo Sinclair Dam, the largest and most controversial project in Ecuadorian history. With respect to the latter, according to investigative journalists, the Ecuadorian government repeatedly failed to enforce its own labor and safety regulations at the behest of the Chinese officials in charge of the project. Thirteen workers died during construction. Even more troubling, this dam was constructed in a seismically active zone near the Reventador volcano, an area that scientists have been insisting for almost 50 years should be off limits to development projects.

After years of spectacular governmental mismanagement, Venezuela has declined to the point that food and medicine supplies are critically short, and corruption, looting, kidnapping, and murder are rife. In a desperate attempt at economic relief, President Maduro has established the Arco Minero ("mining arc") that covers 12% of the country, including enormous swaths of rainforest, numerous indigenous communities, and much of the spectacular Canaima National Park, home to the largest collection of tepuis found within a protected area. Chinese mining companies are believed to have obtained large mining concessions within the Arc.

In Bolivia, employees of the Chinese National Petroleum Corporation, one of the world's biggest oil companies, having been exploring in the northern part of the country and have come close to encountering isolated tribes ill prepared to contact the outside world. Chinese teams are building roads into the rainforest, which is opening up remote areas. Chinese businessmen are also paying local peoples to kill jaguars, whose teeth and testicles are then marketed as aphrodisiacs in China.

Having declared itself socialist after independence from the United Kingdom in the 1960s, Guyana has long enjoyed closer diplomatic relations with China than the rest of South America. In one recent example, a Chinese logging company established operations there seeking enormous rainforest timber concessions. They pledged to hire local workers and build a local wood processing plant. Ultimately, the company was found to have exploited workers by flouting local labor laws, and it did not build the plant. With these breaches of contract and a lack of local economic impact, Guyana forced the company to shut down operations; the Guyana Forestry Commission seized company assets, claiming a debt of $80 million.

In Suriname, Chinese immigrants—Hakka Han from Guangdong province—arrived in the mid-1800s; the nation has had a Chinese-speaking resident population ever since. However, during the past 15 years, there has been a massive influx of Chinese people and businesses into the country. The recent immigrants have established banks, casinos, companies, factories, restaurants, shops, and supermarkets that have driven numerous locals out of business. Suspicions exist in Suriname that this population is a virtual beachhead for Chinese designs on the nation's—and even the continent's—natural resources.

What caused the Amazon fires of 2019?

While Amazonian ecosystems may have always experienced relatively insignificant blazes ignited by lightning, anthropogenic fires in the region began with the arrival of the first humans more than 10,000 years ago. Indigenous peoples in South America set fires in grassland savannas to increase the visibility and abundance of game. These early Amazonians also developed a method of "farming" the rainforest—slash-and-burn agriculture—based on felling small tracts of rainforest, letting the plots dry, and then using controlled burning of the vegetation to both clear the land and release nutrients into the soil to nourish crops.

As mentioned earlier, the first large-scale intentional clearing and burning in tropical South America probably began with Henry Ford's rubber project at Fordlandia in the early 20th century. Even so, the felling of the rainforest with axes was a difficult and time-consuming task.

For many decades, the failure of Fordlandia served as a disincentive for similar grandiose development plans. However, the appearance of inexpensive and efficient chainsaws after World War II made rainforest clearance easier than ever before. And in the 1960s and 1970s, Amazonian countries—led by Brazil—began discussing the need to settle, develop, and integrate the nine countries of Amazonia. Sizeable and ambitious development projects like the previously described Jari Project (1967) and the Trans-Amazon Highway (1972) were initiated. Organizations like the World Bank funded enormous development projects that led to massive deforestation. Many of these efforts spurred additional development of cattle ranching, which moved further into the forest. Though many of the ranches created at this time were by peasant peoples on small plots of lands, several efforts—like the King Ranch and the Volkswagen Cristalino Ranch—were designed to be enormous operations from the outset. And as new techniques and massive inputs of pesticides have made farming in the

Amazon more profitable, particularly at a larger scale when liquidity is unlimited, there has been greater economic incentive to establish more enormous farms.

In October of 2018, former military captain Jair Bolsonaro won the presidency of Brazil after running on a pro-business, nationalist, law-and-order agenda. At the time, Brazil was plagued by skyrocketing crime, a dormant economy, and the largest financial scandal (*lava jato*—"car wash") in the nation's history. Citizens believed that a strong hand was needed to put Brazil on a positive course.

Bolsonaro quickly established himself as the most anti-environmental president in Brazilian history, encouraging his fellow citizens to move into the rainforest and raze it for development. The response to his call to action is believed to have been a major cause of the resultant spike in deforestation and the subsequent massive number of fires. At the same time, he promoted the expansion of extractive industries like mining and logging into Amazonian national parks and indigenous reserves. Bolsonaro also weakened the protection of these lands, slashing funds for enforcement of environmental laws and undercutting the morale, staffing, and budget of scientific agencies. Environmentally minded Brazilians predicted a catastrophe.

The disaster did not take long to manifest. An abnormally high number of large fires at the commencement of the dry season began to alarm both local and foreign scientists. By August, more than 80,000 fires had been counted in Brazil in that year, an increase of more than 80% over the previous year, and the vast majority were in the Amazon. On August 20, 2019, the scale of the crisis was brought home to a good portion of the populace when the smoke emitted by these fires blackened the skies of Brazil's largest city, São Paulo, hundreds of miles away from Amazonia. And the alarming scenes of the rainforest in flames—even if some of the footage was later shown to have been degraded pastureland rather than virgin forest—gripped television audiences around the world, resulting in the hashtag #PrayForAmazonas being the top trending topic on Twitter for a day that same month.

Brazilian government officials angrily denied responsibility and went so far as to blame local NGOs and foreign conservationists for deliberately igniting these fires to cast Brazil in a negative light. They dismissed the chief of the Instituto Nacional de Pesquisas Espacias (INPE), the highly respected Brazilian space agency that tracks forest fires, accusing him of exaggerating the levels of deforestation despite similar conclusions having been reached by NASA, which had also been closely monitoring the situation.

The Brazilian government steadfastly rejected international assistance until the combination of domestic dismay and international pressure proved overwhelming. By late August, the G7, major investment banks, and civil societies around the world were engaged. Several European governments discussed enacting a boycott against Brazilian beef. Ultimately, President Bolsonaro announced the launch of "Operation Green Beef," in which 43,000 military personnel and 2,500 firefighters were sent to battle the blazes. And he announced a 60-day moratorium on Amazon fires.

As the rains commenced in October, the fires began to abate, but much damage had already been done. Less widely reported in the international press was the fact that major infernos had materialized in the rainforests of Bolivia and the dry forests of Paraguay, countries that had even less environmental monitoring and protection than neighboring Brazil. President Evo Morales of Bolivia was notably outspoken in welcoming foreign assistance to help combat the blazes even as he was criticized by some of his constituents for enacting development policies that they believed had made the country more prone to forest fires.

Nonetheless, the impact of these forest burnings on indigenous peoples is still being measured: though less impacted than lands outside protection, tribal lands were also burned. At least 130 different indigenous territories suffered various degrees of conflagrations. Even some of the most remote territories—including 15 forests that are home to isolated and uncontacted peoples—were known to be affected.

Only time—and the next dry season—will give a clear indication if the Brazilian government has retreated from direct and forceful promotion of Amazon deforestation. The governments of all nine Amazonian countries are more conscious of the issue than ever before. Yet no one can be certain that the worst of the fire-driven forest destruction is behind us, especially since climate change continues to accelerate. The success of fighting future fires will depend not only on enforcement of existing environmental laws but also on the funding, training, and equipping of local firefighting forces in Amazonia, along with the establishment of early warning systems to locate and control fires before they burn out of control.

8

CONCLUSION

How can Amazonia be saved?

Protected areas form the backbone of rainforest conservation. Amazonia is home to several colossal national parks: both Chiribiquete in Colombia and Tumucumaque in Brazil are substantially larger than Belgium, while Alto Purus in Peru exceeds the size of New Jersey. These enormous reserves offer the best hope for the protection of relatively intact rainforest ecosystems that harbor healthy populations of apex predators like jaguars and pumas and preferred game species such as spider monkeys, woolly monkeys, and tapirs as well as large birds like harpy eagles and guans. Forest fragmentation in much of Amazonia has demonstrated that sizeable protected areas play a vital role in the survival of species that cannot exist if they need to commute between widely separated smaller parcels of rainforest. As such, protected areas should be regarded as sacrosanct components of the regional, national, and global patrimony.

Another enormous conservation opportunity, one all too often overlooked by the environmental community and the outside world until relatively recently, is represented by indigenous reserves and territories. These homelands total approximately a quarter of Amazonia—about the same amount of land found in the national parks—and the rainforests and rivers in these indigenous areas are often better managed and protected than elsewhere in tropical South America. Tribal peoples living a relatively traditional lifestyle maintain both a physical and a spiritual tie to the forests and the waters. Everything from the quality of drinking water to the availability of foodstuffs to the accessibility of medicinal plants is dependent on careful stewardship of local resources. Many tribal peoples are well aware that, if the forests disappear, they will lose not only their subsistence and economic base, but their cultural identity as well. And not only do Amerindians control enormous amounts of territory, but they have an unparalleled knowledge of these resources and how best to manage them. Ensuring that the collective rights of indigenous communities are respected by local, regional, national, and international forces represents an important step forward.

In Amazonia, large national parks and sizeable indigenous reserves were typically established in remote areas, far from urban centers. This remoteness, however, has made them difficult to monitor and protect. Primarily due to road building, the outside world is pressing into the most remote reaches of the rainforest. Yet the increasing availability and declining costs of satellite imagery mean that surveillance of even the most isolated corners of Amazonia has been facilitated as never before.

With new technology, it is now theoretically possible to monitor every single tree from the air—but monitoring means little without protection and enforcement on the ground and in the rainforest. One obvious solution is to train and employ forest inhabitants to protect their ecosystems: the Amazon Conservation Team has trained and equipped an indigenous park guard force in the northeast Amazon that has been carrying out its duties patrolling the boundaries and protecting the forest for more than a decade. Similar efforts could be launched throughout the Amazon, not just with Amerindians but also with all local communities that will have an increasingly vested interest in wise stewardship of local resources. This effort would integrate the management and protection of protected areas and indigenous territories—around half of Amazonia's standing forest. Moreover, the guards and rangers would be ideally placed to garner income as ecotourism guides as well. Such forces would need to be trained in partnership with local and national governments if they are to be considered national assets due to their ability to report activities that adversely affect the national patrimony.

The failure to provide sufficient economic benefits to the rural poor remains a major driver of deforestation in the Amazon. These marginalized peoples are continuously driven into more remote forests or even protected areas because prosperous landholders employ gunmen to guard their own property or seize those of others, often well beyond the reach of the law. Therefore, assisting poor famers in gaining title to their lands represents a means to combat the destructive cycle of endless migration and deforestation. In the words of rainforest journalist Rhett Butler: "Once local people have a stake in the land they are farming, they have an interest in using it efficiently instead of moving on to a new area of forest once soils are prematurely exhausted."

Such programs have been successfully implemented recently in both Colombia and Peru. At the same time, however, complementary efforts are necessary to increase the likelihood of success over the long term. More effective law enforcement is essential if small stakeholders are to effectively fend off predatory efforts by wealthier and more powerful landholders and corporations. In fact, combating any and all forest crime—from murders of environmental activists to exploitation of the poor (including actual slavery as documented in Brazil) to illegal logging to narcotrafficking—needs to be made a much higher priority by the governments of all nine Amazonian countries.

Yet another way to provide a helping hand to forest communities is through the provision of access to credit and other banking functions, but only in ways that encourage good environmental stewardship and do not finance additional deforestation. Being able to both borrow and save money can transform the way people are able to take control of their financial and environmental destinies—the microfinancing efforts of Bangladesh's Grameen Bank have revolutionized thinking about how best to assist marginalized populations while unleashing entrepreneurial potential.

Meanwhile, the global business community—concerned by almost ubiquitous indications of the accelerating rate of climate change—is employing more of its influence to slow the destruction. In 2019, with people marching in the streets demanding action to protect forests while tens of thousands of fires raged in Amazonia, a group of more than 230 global investment funds—managing more than $16 *trillion*—signed a letter asking companies to implement policies designed to combat deforestation in Amazonia.

Simultaneously, marginalized peoples will need additional training in improving agricultural techniques. Permaculture techniques that rely more heavily on planting tree crops than is common in traditional Amazonian agriculture is one such example; such systems are less destructive to both the soil and to plants and animals. Another method of assisting these farmers is by introducing ancient practices such as the production of "terra preta" soils that are both more resilient and more productive than typical Amazonian soils.[1] Their production is further strongly warranted by the added advantage of carbon sequestration offered by these soils. And other pre-Columbian techniques of agroforestry, fisheries management, irrigation, and raised field agriculture—now being studied in the Llanos de Moxos in the Bolivian Amazon—could provide new insights into better stewardship of tropical resources.

Large-scale agricultural and industrial operations (primarily forestry, fishing, and mining) have played an outsized role in the deforestation of the Amazon. By their nature, these enterprises are inimical to biodiversity. Nonetheless, there are ways that they can mitigate their direct impacts and contribute to the overall protection of Amazonia. An obvious first step is the reduction or elimination of the release of toxic and destructive chemicals into the environment, such as the cyanide and mercury employed in gold mining as well as the harmful pesticides and fertilizers used in industrial agriculture.

More and better research is needed on the biology of commercial species of plants and animals, particularly large-scale harvesting of fish and high-value timber. In a recent study of the legal wildlife trade in Amazonia by the United Nations Environmental Program,[2] the value of parrots and turtles for the pet trade and caiman and peccary skin for the fashion industry was estimated to be $128 million annually. Research into captive breeding of these and other species, as well as cracking down on illegal trade—such as the recent but lamentable Chinese trade in "medicinal" jaguar teeth—could further benefit rural peoples while relieving some of the pressure on the animal populations.

Underutilized but economically promising species of plants—like cacay (*Caryodendron*), which thrives in degraded soils and produces a high-quality oil for cosmetics; tabakabon (*Cordia*), a rapidly growing tree that also thrives in poor soils and yields high-grade firewood; or even the better-known tagua palm, capable of producing greater quantities of vegetable ivory—merit further research. Sustainable timber harvest has proved to be more difficult to conduct than once envisioned, and timber certification efforts have been plagued with abuse—though improvements in technology have the potential to significantly increase integrity and reliability. Reduced-impact logging (RIL) and detailed land use planning—for forestry and all extractive industries—would represent a major step forward. Governments can deem high-diversity areas off limits to development and offer incentives for easements in areas that have a high conservation priority.

Meanwhile, though efforts to ensure that cattle, soybeans, palm oil, and other products are produced relatively sustainably without further deforestation are becoming more verifiable, they are not yet where they need to be to secure environmental protection. In any case, large agricultural and industrial enterprises that are involved in destructive activities should be required to finance environmental set-asides: better protection of existing protected areas or the creation of new ones.

Recovery and rehabilitation of degraded lands offer great potential to produce cattle, crops, and timber while relieving pressure on pristine rainforests. As scientists gain a better understanding of tropical ecosystems—especially soils—areas that have been regarded as wastelands could theoretically be brought back into production. A research focus on enhancing the utility of existing pastures and plantations could meet growing demand for land. EMBRAPA, the Brazilian government's agricultural research agency, has stated that farming in the Amazon can be improved and upgraded through more creative use of lands considered degraded and otherwise of little commercial potential.

A proven way to generate income for multiple levels of Amazonian societies is through improved ecotourism. Tourism represents one of the world's largest industries, and demand continues to grow. Amazonian countries often look to Costa Rica, where ecotourism serves as the main source of foreign exchange. Perhaps a better model to follow might be the mountain gorilla conservation efforts in Rwanda, in that the focus is on attracting high-end, low-impact ecotourists; rainforest conservation does not lend itself to the tourist hordes that throng the savannas of East Africa. Promoting so-called *niche tourism* to wealthy birdwatchers or sport fishermen represents one already functioning model for attracting top dollars from a relatively small clientele that has a clear interest in enjoying a well-preserved and managed ecosystem.

Of course, mass ecotourism—particularly in the tropics—can and does have deleterious effects and must be managed carefully, not only to minimize impacts on the environment but also to ensure that local communities benefit, thereby creating an economic incentive to protect the Amazonian environment. All too often, monies generated in the rainforest end up mostly in the hands of rich and powerful urban elites—even in Costa Rica.

Creative financial mechanisms offer yet another means of generating monies for proper rainforest stewardship. In 1984, Thomas Lovejoy of the World Wildlife Fund proposed what he termed *debt-for-nature swaps*, in which international non-governmental organizations (NGOs; usually conservation organizations) purchase the commercial debt of tropical countries from creditors at a deep discount while agreeing with the countries on repayment schedules. The money thus earned by the NGO is then employed to finance in-country conservation activities. Alternatively, deals may be struck where, in exchange for a cancelled debt note, the country agrees to enact certain environmental measures. The first transaction of this kind was carried out by Conservation International in 1987, when the organization purchased $650,000 of Bolivian debt for $100,000 and earmarked the monies for environmental protection in the Bolivian Amazon. Subsequent swaps resulted in more than a billion dollars being funneled into conservation activities throughout the tropics, including Ecuador and Venezuela.

Another innovative scheme is *payment for ecosystem services* (PES), in which direct payments are made to landowners who protect forests whose benefits accrue to others. The services produced by this forest maintenance are biodiversity protection (notably crop pollinators), carbon sequestration, hydrological services (especially generation of rain, water filtration, and regulation of water flows), and preservation of scenic beauty (a boon to ecotourism). This approach is being extensively applied in Costa Rica.

A more widely discussed methodology for generating rainforest conservation funds is *Reducing Emissions from Deforestation and Forest Degradation Plus* (REDD+) conservation, sustainable management, and enhancement of forest carbon stocks. Essentially, this is a program in which tropical countries are paid to protect their forests and thereby cut greenhouse gases. However, indigenous communities have justifiably complained that some early efforts failed to include them in both the decision-making process and the revenue stream. In response, they created the Amazon Indigenous REDD+ Initiative (RIA in Spanish), which is now being implemented in Brazil, Colombia, Ecuador, and Peru. Although a lack of funding for the REDD approach limited its effectiveness, REDD did help catalyze important discussions about recognizing indigenous land rights.

Better and more accurate economic analysis can also play a key role in promoting sustainable development. Multilateral organizations like the World Bank and the Inter-American Development Bank long predicated their financing on an economic rate of return that downplayed or even ignored environmental and social costs. A review of their economic analyses reveals that they and the host governments repeatedly underestimated the destructive effects of rainforest dams and typically overestimated the benefits. Their penchant for funding giant infrastructure projects was a function of their desire to administer as few projects as possible. And their inability to understand and put a dollar value on the aesthetic, biological, climatological, ecotouristic, ethnomedical, geochemical, and hydrological benefits of rainforests consistently resulted in their underestimating the value of Amazonian ecosystems. Attitudes are changing, but mega-infrastructure projects—particularly road building—are still under way. The good news is that multilateral development banks like the International Development Bank (IDB) and the World Bank have increasingly realized the downside of these massive infrastructure efforts; the bad news is that other institutions, like national development banks and the Chinese, are more likely to fund these efforts. On this and several other environmental fronts, the Chinese can and will cast deciding votes on whether economic "development" is sustainable or massively destructive.

Nonetheless, the abundance of data now available to decision-makers has never been greater: whether through satellite imagery, biological surveys, or learning from past mistakes, development planners can make much better decisions. One example is that of Professor William Laurance and his colleagues, who have designed "smart" road-building plans that minimize destruction of biodiversity and stream connectivity; these models demonstrate that conservation and economic development need not always work at cross purposes.

One urgent priority must be increased appreciation for and protection of freshwater ecosystems. Fish remains the major protein source for most Amazonian populations outside of large cities, and infrastructure projects that generate deleterious impacts on fish and potable water must be mitigated or avoided outright. We still know precious little about the life histories and ecologies of most Amazonian aquatic species. One of the best known—the piraiba, or giant catfish (*Bradyplatystoma filamentosa*)—lives, thrives, and spawns from the mouth of the Amazon to the foot of the Andes, a distance of more than 3,000 miles (4,828 km). This means not only that more research must be conducted to better understand these species and the function and interdependence of the ecosystem, but also that management schemes need to be geared toward the Amazon as a whole, rather than just using a regional or just a national approach. Connectivity of Amazonian rivers and other freshwater systems—threatened by the continued construction of huge dams—must be maintained, with an especially high priority given to headwaters and headwater forests from which the rivers originate.

Heartening is an increasing desire on the part of business—particularly big business—to be seen as environmentally friendly or at least not environmentally destructive. Consumer movements that led to the Soy Moratorium in Brazil and the Rainforest Beef Boycott in Central America were vivid demonstrations of consumer muscle in terms of demonstrating how purchasing power can punish environmentally destructive companies. Major global corporations like Unilever have announced "Zero Deforestation" pledges that represent steps in the right direction. But more must be done, certainly in terms of both agricultural products and timber: Daniel Nepstad proposes that companies partner closely with countries and communities not only to ensure that positive environmental stewardship becomes more common but also to provide additional opportunities and training to spread benefits even more widely.

Energy development must be part of the discussion as well: because Amazonia receives as much sunlight as any other region on the planet, harnessing solar power must be a top priority. Solar panels are increasingly common in even some of the most remote regions in Amazonia. The ever-declining cost and ever-increasing output of solar technology means that this power source will prove increasingly attractive. And the ugly record of oil production and pollution in the western Amazon provides strong incentives to South American countries to increase reliance on solar while experimenting with wind power and other renewable energy sources.

Fostering climate resilience also must be a priority goal. Deforestation has already disrupted and diminished rainfall patterns to the detriment of everyone from indigenous farmers to large South American cities. At this point, what may be required is an Amazon-wide reforestation policy that is revolutionary in scale—for example, for every tree that is cut, two must be planted. Also necessary is protected-area design and management with an eye toward mitigating climate change, whether by increasing connectivity between isolated forest stands or by ensuring that protected areas are as large as possible.

Finally, protection of the Amazon will require a different mindset. If the driving force behind all decisions that affect the great rainforest are merely maximizing short-term economic return to global elites, the Amazon is doomed. The ardently anti-environmental agenda of the current Brazilian government—echoed by that of the US government—provides ample reason for pessimism.

If development is only measured in terms of short-term greed, there is no hope for the proper stewardship of this or any other rainforest ecosystem. However, if we can modify our goals and our approach to encourage sustainable harvest and production and long-term planning while incorporating societal well-being—embracing clean air, clean water, the deceleration of climate change, the promotion of equitable opportunity for all, and human rights, particularly for marginalized populations—then there is most definitely hope.

When I was growing up in the 1950s and 1960s, people habitually threw litter out their car windows, smoked cigarettes in offices and airplanes, shunned seatbelts, and assumed the Berlin Wall would never come down. With enough changed minds come changed policies and realities. With respect to the Amazon, we await that critical change.

NOTES

Chapter 1

1. The term "Guayana" or "Guyana" or "Guiana" is often described as being derived from an Amerindian word meaning "land of waters." The dominant Amerindian language group in northeastern Amazonia—the Guianas—is Carib. In Carib languages, the suffix "yana" means "people"—e.g., "Tunayana" means "water people," and "Okomoyana" means "wasp people." One of the largest groups are the "Wayanas"—"people of the wy tree." The first explorers of the Guiana coast were Spaniards—Ojeda and de la Cosa—in 1494, just 2 years after Columbus. Hearing of "Wayanas" who lived in the interior, a Spaniard would record on his map the name "Guayanas," which is the origin of the term.

Chapter 2

1. Terra preta contains more than 60 times the amount of charcoal of nearby soil, and the charcoal is in the form of *biochar*, which is produced by heating organic matter at a relatively low temperature in a low-oxygen setting.
2. Plants are not the only rainforest denizens seeking to find and filch nutrients. Observant visitors can witness dung beetles rolling excrement deposited by sizeable creatures (including tourists) into balls that they then trundle to their nests so their larvae can feast on the nutrients within. Colorful Trigonid sweat bees in search of salt can be persistent nuisances to the sweaty rainforest tourist. And the next time you see an Amazon film featuring graceful green and yellow Pieridae butterflies alighting

onto a sandy riverbank, know that this scene was likely set up by means of an old cinematographer's trick: urinating on the sand prior to filming, thereby luring the butterflies to swoop down en masse and drink the salts within.

3. In a classic study in Venezuela, soil scientist Carl Jordan found that 99.9% of the calcium and phosphorus released into the ecosystem was adsorbed (attached) to the root mat by the mycorrhizae. Jordan, C. F. (1982). "The nutrient balance of an Amazonian rain forest." *Ecology*, 63: 647–654.
4. Daly, D. C. & J. D. Mitchell. Lowland vegetation of tropical South America—an overview. Pages 391–454. In D. Lentz, ed. *Imperfect Balance: Landscape Transformations in the pre-Columbian Americas*, (Columbia University Press, New York, 2000).
5. Another isolated type of hill in Amazonia sometimes mistaken for tepuis are *inselbergs*. However, tepuis are flat-topped sandstone mesas while inselbergs tend to be rounded domes of granite. Like tepuis, inselbergs are found from Colombia to the Guianas, but the most famous inselberg in South America lies outside of Amazonia: Rio de Janeiro's iconic Pão de Açucar, known in English as Sugarloaf Mountain.
6. Im Thurn, Everard. (1885). "The ascent of Mount Roraima." *Proceedings of the Royal Geographical Society and Monthly Record of Geography*, 7(8): 497–521, p. 517.

Chapter 3

1. The first paddle steamers—the vessels that brought the industrialized world to Amazonia—appeared in the early 1850s and traveled as far upstream as Nauta, Peru, approximately 60 miles (96 km) above the location that would later become the city of Iquitos. Smaller riverboats can travel up to the Pongo de Manseriche, another 420 miles (675 km) beyond Nauta.
2. Though some have suggested that the National Geographic representatives were the first people to reach this forbidding and almost inaccessible locale, McIntyre told me that he and his team had noted the presence of Inca mummies, indicating that not only had Indians been there before them but that they had worshipped it as a sacred site.
3. Of course, Wallace "discovered" these river types like Columbus "discovered" America—in both cases, the Indians got there first.

In fact, many great scientific "discoveries" have consisted of writing down what the local peoples already know.

4. Beginning in the 1950s, the German-Brazilian limnologist Harold Sioli provided additional hydrochemical and physical characteristics to further support Wallace's categories. Sioli was also a pioneer in studying the Amazon from a basin-wide perspective, a tireless and effective campaigner for protection of the Amazon, and a Cassandra-like figure who warned that rainforest destruction would result in a massive increase in atmospheric carbon.
5. The origin and meaning of the name "Casiquiare" remain unclear. Linguist Alexandra Aikhenvald notes that the term "ari" in Amazonia often means "river" or "waterway." And a common term in tropical South America for chief is "cacique." Thus I hypothesize that "Casiquiare" means "The River of the Chief."
6. Rhodoliths are colorful red algae that resemble coral.

Chapter 4

1. These artifacts indicate the use of plants of the *Anadenanthera* genus, which contain a powerful hallucinogen still widely employed in the northwest Amazon, as well as other psychoactive plants like coca and tobacco.
2. Schultes, R. E., and A. Hoffman. *Plants of the Gods* (New York: McGraw Hill), 1979, p. 120.
3. The name of the genus has since been changed to *Banisteriopsis*.
4. Subsequent research revealed that telepathine was identical to harmine, an alkaloid that had previously been isolated from Syrian rue. Israeli ethnobotanist Benny Shannon noted that the active compounds in the ayahuasca potion—both harmine/telepathine in the Syrian rue (*Peganum harmala*) and dimethyltryptamine (DMT) from the Acacia trees (*Acacia* spp.)—are widespread in the Sinai and hypothesized that Moses' ingestion of these compounds could have produced visions of burning bushes and encounters with the Divine.
5. Ethnopharmacologist Dennis McKenna—the leading authority on the chemistry of ayahuasca—notes that even much-studied alkaloids can yield novel findings. Harmine was isolated in 1847, but recent research has revealed that harmine and some of its derivatives exhibit anticancer, antidepressant, antidiabetic, and antimicrobial properties.

6. Schultes and Hoffman, op. cit., p. 7.
7. One species of bromeliad—*Pitcairnia feliciana*—is found in West Africa, but it may have been an accidental introduction from tropical America.
8. The technical term for a water-filled cavity in a plant is "phytotelma," and the plural is "phytotelmata."
9. Armbruster, P., R. Hutchinson, and P. Congreave. (2002). "Factors influencing community structure in a South American bromeliad fauna." *Oikos*, 96: 225–234.
10. Schultes, R. E. (1974). "Palms and religion in the Northwest Amazon." *Principes*, 18(1): 3–21, p. 3.
11. Ter Steege, H., R. Vaessen, D. Cardenas-Lopez, et al. (2016). "The discovery of the Amazonian tree flora with an updated checklist of all known tree taxa." *Scientific Reports*, 6: 29549. http//doi.org/10.1038/srep29549.
12. Haenke is one of history's greatest unsung natural history explorers, studying and exploring from Alaska to Acapulco, from the Amazon to the Andes, and from Australia to New Zealand to Tonga. He made these treks from 1789 to 1816, at which point he was accidentally poisoned by his cook before he was able to record all his observations and conclusions for posterity. Typically, the great naturalists of the 19th century, like Charles Darwin and Henry Walter Bates, published their masterworks long after they had completed their tropical fieldwork.
13. Over 7 feet (2 m) in length and weighing more than 100 pounds (45 kg), the Amazonian giant anteater (*Myrmecophaga tridactyla*) represents the largest living descendant of South America's early mammals.
14. Once again, the lack of competition fostered the rise of giant species: though today the 200-pound (90 kg) capybara is the world's largest rodent, one extinct caviomorph weighed over half a ton (450 kg).
15. All New World monkeys, including the 14 genera in the Amazon, are believed to have descending from a single group of African arrivals. New World monkeys—known as platyrrhines—tend to be arboreal, have prehensile tails, and have nostrils that are widely separated and open to the side. Old World primates (including humans) are technically known as catarrhines, lack tails, and have nostrils that point downward. The largest Amazonian primate is the wooly monkey, which can weigh

more than 20 pounds (9 kg), while the smallest is the pygmy marmoset, which tips the scale at just over 3 ounces (85 g).

16. That this interchange had taken place was perhaps first remarked upon by Alfred Russel Wallace in 1876, 24 years after his departure from the Amazon.
17. Vampire bats are not the only Amazonian creatures being studied as a potential source of new anticoagulants. The giant Amazon leech (*Haementeria ghilianii*) reaches a length of 18 inches (45 cm) and sucks blood as a source of sustenance. This leech produces a different coagulant (hirudin) than does the bat, hence the leech extract has the potential to serve as a therapy for blood maladies that draculin can treat, as well as others it cannot. Given that there are close to 800,000 cases of stroke each year in the United States alone—and that people are living longer (meaning the incidence of stroke is expected to rise)—the search for new stroke treatments in both the rainforest and the lab will continue.
18. Bates, H. W. (1988). *The Naturalist on the River Amazons* (New York: Penguin), p. 80.
19. Because of this peculiar and unforgettable pose—with two hairy arms held up in the air—the Brazilians call this creature "aranha macaco": the spider monkey spider.
20. In fact, novel poisons can be important to medicine in several ways. At a lower dose, a toxin can sometimes be curative. Or, a poison can be manipulated chemically and converted into a therapeutic compound. And poisons can be studied to gain a better understanding of human form and function, particularly with regard to the nervous system. I explain this in greater detail in my 2001 book, *Medicine Quest* (New York: Penguin).
21. The order Crocodilia contains two families in the New World: the *Crocodylidae* (the true crocodiles like the Orinoco crocodile and the American crocodile) and the *Alligatoridae* (the alligators and the caimans). Thus, a black caiman is a crocodilian, but not a crocodile.
22. Bates, op. cit., p. 295.
23. Boas, anacondas, and the more distantly related pythons are considered primitive snakes. On either side of the cloaca, they have "pelvic spurs," vestigial remnants of the legs possessed by their ancestors.
24. Bates, op. cit., p. 211.

25. Scientists have not reached a consensus as to why the botos are pink. Theories include that the color provides camouflage (to blend in with the red mud) or is caused by capillary placement because of scarring, due to males battling for dominance.
26. Arowanas are so-called mouth brooders: the male transports and protects the tiny fish fry in his mouth for about 6 weeks after they hatch.
27. Lundberg, J. (September 2001). "Freshwater riches of the Amazon." *Natural History*, 36–43, p. 42.
28. China is not much larger than the United States but it has more than three times the botanical diversity of North America. That China was not subjected to Pleistocene glaciation explains in part why China is so much richer in terms of plant diversity.
29. These cetopsid candirus also are said to enter and/or bite vaginas of women bathing in rivers. Some go so far as to blame them for breaking a young girl's hymen, thereby annulling her virginity. T An Amazon ichthyologist once recounted the following to me: "A *caboclo* [local peasant] on the Rio Madeira once told me that he had a priest in Porto Velho declare his daughter a virgin despite no longer having a hymen because it was a candiru that had broken it. I suspect that the candiru's name was Pedro or José!"
30. Goulding, M. (1990). *Amazon: The Flooded Forest* (New York: Sterling Publishing), p. 136.
31. Sharks are cartilaginous fish; piranhas are bony fish.

Chapter 5

1. Yukuna Indian longhouses (*malocas*) in the Colombian Amazon can reach three stories in height and are unfailingly watertight in even the most unremitting and torrential tropical downpours. And—like all traditional dwellings—they are constructed without the use of a single nail.
2. The Trio Indians insist that their ancestors in the distant past crossed a land that was so cold that they had to wrap themselves in the skins of animals. Indians in northern Argentina relate a similar tale.
3. The lowest pass through the Andes is the Abra de Porculla in northern Peru. At just over 7,000 feet (2,134 meters), it would not have presented an insurmountable obstacle for those bold and hardy enough to make the trek from Asia. In Colombia,

the lowest Pacific-to-Amazon transit is about 9,000 feet (2,743 meters), near the Sibundoy Valley, which the local Indians claim has been inhabited since ancient times and which is relatively near Pena Roja, the second oldest known site in Amazonia.

4. Mora, S. (2003). *Early Inhabitants of the Amazonian Tropical Rain Forest* (Pittsburgh: University of Pittsburgh Latin American Publications), p. 7.
5. Barlow, J. T. Gardner, A. Lees, L. Parry, and C. Peres. (2011). How pristine are tropical forests? *Biological Conservation*, 151: 45–49, p. 48.
6. In 1743, the French explorer Charles Marie de La Condamine descended the Amazon from Ecuador less than two centuries after Orellana and was stunned to sail for days along uninhabited riverbanks once crowded with huge villages when the Spanish conquistador made his epic voyage.
7. On the difference between *pidgin* and *creole* languages, since both Lingua Geral and Sranan Tongo are creoles: according to linguist Alexandra Aikhenvald, "A creole is a language that historically develops from a pidgin. A pidgin develops from trade or other contact; it has no native speakers, its range of use and vocabulary are limited, and its structure is simplified. It later becomes the only form of speech common to a community, is learned by new speakers, acquires native speakers, and is used for all purposes; its structure and vocabulary become more complex. At this stage, it becomes a creole."
8. Castner, J. (2002). *Shrunken Heads* (Gainesville, FL: Feline Press), p. 20.
9. The original range of chile peppers appears to extend from South America into Central America and even the American Southwest. Several Amazonian plants, like cacao, peach palm, and pineapple, were already being cultivated in Central America by the time the Europeans arrived.
10. Bitter cassava seems more prevalent in Amazonia as the Amerindians claim that this variety is more resistant to insects and less appealing to crop-raiding mammals that frequent indigenous gardens.

Chapter 6

1. Prior to assuming the papacy, Pope Alexander VI was known as Rodrigo Lanzol Borgia. He had been born as Rodrigo de Borja: in fact, the uber-Italian Borgias originated as Borjas from Spain.
2. The author has made both journeys. It is much easier to sail or paddle down the Amazon than up the Rio Negro.
3. It stands as a historical irony that Fritz traveled east down the Amazon in search of a cure for malaria when the most efficacious treatment was due west: quinine, native to the Andes.
4. Schiebinger, L. (2008). "Exotic abortifacients and lost knowledge." *The Lancet*, 371(9614): 718–719, p. 718.
5. In their highly acclaimed biography, *Darwin's Sacred Cause*, Adrian Desmond and James Moore propose that Darwin's fervent antislavery beliefs contributed to the formulation of his evolutionary theories. Certainly, the members of the Wedgwood side of his family were fervent abolitionists, so he had exposure to these beliefs early in life. But his time with—and relationship to—John Edmonstone undoubtedly played a crucial role as well. Few upper-class Englishmen in the early 19th century numbered a freed slave among their friends, much less one serving as a teacher and mentor. And the largely nonviolent but bloodily suppressed slave rebellion in British Guiana just 2 years before their collaboration must have entered into their discussions. Darwin was revolted and horrified by the brutal and inhuman treatment of slaves he witnessed in Brazil. "How often Darwin must have seen the amiable John Edmonstone . . . [in the faces of] these oppressed peoples," wrote Desmond and Moore. They conclude that the bond he formed with Edmonstone led him to conclude that all races originated from a common ancestor, one of the cornerstones of evolutionary theory.
6. In 1845, Darwin actually visited Waterton at his ancestral estate of Walton Hall in Yorkshire to meet the famous naturalist who had taught Darwin's teacher. At that time, Waterton was the equivalent of a rock star, having published his bestselling account of his Amazon rainforest adventures as *Wanderings in South America* in 1828, making him one of the most famous and certainly the most colorful naturalists of his day. Darwin, on the other hand, would not produce his masterwork for another 14 years, at which point his fame would eclipse that of Waterton

and virtually all other biologists. Darwin recorded the encounter in a letter sent to his friend Charles Lyell.

> Visited Waterton at Walton Hall, and was extremely amused by my visit there. He is an amusing strange fellow; at our early dinner our party consisted of two Catholic priests and two Mulatresses! He is past sixty years old, [yet] the day before ran down and caught a [small rabbit] in a turnip field. It is a fine old house and the lake swarms with wild fowl. (Darwin, Francis, ed., *The Life and Letters of Charles Darwin, including an Autobiographical Chapter* [New York: D. Appleton and Company, 1888], vol. 1, 343–344).

7. So xenophobic were the Portuguese in Brazil that even the venerable Captain Cook had his men imprisoned, his papers questioned, and his well-being threatened when he docked at Rio de Janeiro in 1768—his ship was quickly provisioned, and he was dispatched. Von Humboldt sailed up the Orinoco and the Casiquiare Canal in Venezuela, up to the Brazilian border, but was turned back in 1800. Charles Waterton walked into Brazil from British Guiana in 1824. Arriving at Fort São Joaquim on the border almost dead from malaria, he was treated and then sent back to British Guiana.
8. Scientists seldom honor indigenous cultures by incorporating local names into scientific names. Fusée-Aublet was an exception: the rubber genus *Hevea* is based on the local name "Eh-way." The Swedish biologist Carl Linnaeus, the father of modern taxonomy, also engaged in this practice: the scientific name he created for the Amazonian howler monkey—*Alouatta*—is the species' name in the language of Suriname's Wayana Indians.
9. In 1735, La Condamine had been sent to the Andes as part of an expedition to determine the precise shape of the earth. Though often overlooked in the annals of ethnobotany, his 10-year sojourn in tropical America can be considered one of the most productive expeditions ever conducted in that his were among the first scientific observations and collections of quinine, curare, and rubber.
10. When chilled, rubber retains its elasticity only down to a certain temperature: the space shuttle Challenger disaster has been attributed to the failure of rubber O-rings to retain their seal when subjected to extreme cold.

11. This system—which still survives today in remote stretches of Amazonia—most commonly involved peasants or mestizos.
12. As a partial act of atonement, in 1988, Colombian president Virgilio Barco established a 21,000-square-mile indigenous reserve to protect the Indians and their forests in an area once ruled by the iron hand of Arana.
13. A sense of the utter destruction caused by the rubber trade was conveyed by the great German anthropologist Theodor Koch-Grünberg: "Hardly five years have gone by since my last visit. . . . Whoever comes here now will no longer find the pleasant place I once knew. The pestilential stench of a pseudocivilization has fallen on the brown people who have no rights. Like a swarm of annihilating grasshoppers, the inhuman gang of rubber barons continues to press forward. The Colombians have already settled in at the mouth of the Kuduyari and carry off my friends to the death-dealing rubber forests. Raw brutality, mistreatment and murder are the order of the day. . . . The Indian villages are desolate, their homes have been reduced to ashes and their gardens, deprived of hands to care for them, are taken over by the jungle." (Koch-Grunberg, T. 1909. Zwei Jahre unter den Indianern. [Berlin: Wasmuth], cited in Schultes, R.E. and R. Raffauf. 1992. Vine of the Soul. [Santa Fe: Synergetic Press], p. 10).

Chapter 7

1. Barbosa, L. (2015). *Guardian of the Brazilian Amazon Rainforest* (London: Routledge), p. 101.
2. Ansar, A., B. Flyvbjerg, A. Budzier, and D. Lunn. (2014). "Should we build more large dams?" *Energy Policy*, 69: 43–56, p. 50.
3. Fearnside, P. (2006). "Dams in the Amazon: Belo Monte and Brazil's Hydroelectric Development of the Xingu River Basin." *Environmental Management*, 38(1): 16–27.
4. Particularly chemical lime, which changes soil pH and eliminates aluminum toxicity.
5. Greenpeace. (2006). "Eating Up the Amazon" (Brasilia: Greenpeace), p. 64.
6. Asner, G. and E. Broadbent, P. Oliveira, M. Keller, D. Knapp, and J. Silva. (August 2006). "Condition and fate of logged forests in the Brazilian Amazon" *PNAS*, 103 (34): 12947–12950. https://doi.org/10.1073/pnas.0604093103

7. To present a sense of scale: annual timber harvest in French Guiana is approximately 3,531,467 cubic feet (100,000 cubic meters); annual harvest in the Brazilian Amazon is approximately 1,236,000,000 cubic feet (35 million cubic meters).
8. Personal communication to the author.
9. Macedo, M., and L. Castello. (April 2015). *State of the Amazon* (Brasilia: World Wildlife Fund), p. 29.
10. Bates, H. W. (1988). *The Naturalist on the River Amazons* (New York: Penguin), p. 292.
11. Ecologist Dan Janzen was even pithier, labeling these forests and rivers "The Living Dead."
12. Redford, K. (June 1992). "The Empty Forest." *Bioscience*, 42(6): 412–422, p. 412.

Chapter 8

1. Traditional creation of terra preta soils required decades, if not centuries. Efforts are under way in Brazil and elsewhere to produce *terra preta nova* (new terra preta) soils with the same attributes that can be generated much more rapidly than the original product.
2. Sinovas, P., B. Price, E. King, A. Hinsley, and A. Pavitt. (2017). *Wildlife Trade in the Amazon Countries,* (Cambridge: World Conservation Monitoring Centre), p. 112.

INDEX

Note: *For the benefit of digital users, indexed terms that span two pages (e.g., 52–53) may, on occasion, appear on only one of those pages.*